Analysis of Cryogenic Propellant Tank Pressurization
based upon Experiments and Numerical Simulations

Vom Fachbereich Produktionstechnik

der

UNIVERSITÄT BREMEN

zur Erlangung des Grades

Doktor-Ingenieurin

genehmigte

Dissertation

von

Dipl.-Ing. Carina Ludwig

Gutachter:

Prof. Dr.-Ing. Michael Dreyer

Prof. Dr.-Ing. Andreas Rittweger

Tag der mündlichen Prüfung:

05.08.2014

Bibliografische Information der Deutschen Nationalbibliothek

Die Deutsche Nationalbibliothek verzeichnet diese Publikation in der Deutschen
Nationalbibliografie; detaillierte bibliografische Daten sind im Internet über
http://dnb.d-nb.de abrufbar.

1. Aufl. - Göttingen: Cuvillier, 2014

Zugl.: Bremen, Univ., Diss., 2014

© CUVILLIER VERLAG, Göttingen 2014

Nonnenstieg 8, 37075 Göttingen

Telefon: 0551-54724-0

Telefax: 0551-54724-21

www.cuvillier.de

1. Auflage, 2014

Gedruckt auf umweltfreundlichem, säurefreiem Papier aus nachhaltiger Forstwirtschaft

ISBN 978-3-95404-797-0

eISBN 978-3-7369-4797-9

Abstract

Focus of this study is the pressurization system of a cryogenic liquid propellant tank for the launcher application. Objective of the pressurization system is to provide the required pressure evolution in the propellant tank for a proper operation of the engine. The motivation of this study is an improved understanding of the complex fluid-dynamic and thermodynamic phenomena during the active-pressurization process of cryogenic propellants in order to optimize the on-board pressurant gas mass. Therefore ground experiments, numerical simulations and analytical considerations were performed for the investigation of the initial active-pressurization phase. For the performed experiments, liquid nitrogen was used as cryogenic model propellant, which was actively pressurized under normal gravity conditions up to different final tank pressures. As pressurant gases, gaseous nitrogen and gaseous helium were used with different inlet temperatures. For the numerical analyses the commercial CFD program Flow-3D, a three-dimensional Navier-Stokes equation solver, was used.

As main results, it can be summarized that the increase in saturation temperature at the free surface during the pressurization phase is identified as one main driver of the establishment of a thermal stratified layer below the liquid surface. It is moreover found that in order to minimize the required pressurant gas mass, for the performed experiments a high pressurant gas temperature and/or the application of helium as pressurant gas is found to be advantageous. During the active-pressurization with gaseous nitrogen, condensation is identified as the predominating mode of phase change and with helium as pressurant gas, evaporation dominates. After pressurization end, only condensation appears. Furthermore, a "smart fit" equation is derived, which quite accurately describes the evolution of the pressure drop after pressurization for the performed nitrogen pressurized experiments. Based on this equation, the decrease of the vapor temperature after pressurization end is identified as the main driver of the pressure drop. As the dominating heat transfer during pressurization, the heat transfer from the injected pressurant gas to the axial tank wall is confirmed, which is therefore dependent on the pressurant gas temperature. This heat is conducted inside the wall toward the liquid phase, contributing to the temperature increase in the area nearby the tank wall of the upper liquid layers. By considering the heat transfer from pressurant gas to the tank wall and disregarding the influence of phase change, a correlation is established which allows an a priori determination of the tank pressure rise of the performed experiments. Moreover, based on the JAKOB number and the thermal expansion FROUDE number, a correlation is established for the determination of the required pressurant gas mass. This study describes in detail the performed active-pressurization experiments, the numerical model and analytical approaches and correlations. The conclusions drawn from these results are established and discussed.

Zusammenfassung

Die vorliegende Arbeit beschäftigt sich mit Untersuchungen zur Bedrückung von flüssigen, kryogenen Treibstoffen in Raketentanks. Aufgabe des Bedrückungssystems ist es, den erforderlichen Druckverlauf in den Tanks für den reibungslosen Betrieb der Treibwerke zu gewährleisten. Motivation dieser Arbeit ist ein verbessertes Verständnis der komplexen fluid-dynamischen und thermodynamischen Phänomene, die während der aktiven Bedrückung kryogener Treibstoffe auftreten, vor dem Hintergrund die on-board Druckgasmasse von Raketen zu optimieren. Im Rahmen dieser Arbeit wurden Bodenexperimente und numerische Simulationen durchgeführt sowie analytische Ansätze angewandt. Der Fokus lag dabei auf der ersten aktiven Bedrückungsphase. Für die im Rahmen dieser Arbeit durchgeführten Experimente wurde flüssiger Stickstoff als kryogener Modelltreibstoff verwendet, der aktiv auf verschiedene Tankenddrücke bedrückt wurde. Als Druckgase wurden gasförmiger Stickstoff und gasförmiges Helium unterschiedlicher Temperaturen verwendet. Für die numerischen Analysen wurde das kommerzielle CFD-Programm Flow-3D, ein dreidimensionaler Navier-Stokes Gleichungslöser, verwendet.

Als wichtigste Ergebnisse kann zusammengefasst werden, dass der Anstieg in der Sättigungstemperatur an der freien Oberfläche während der Bedrückungsphase als ein maßgeblicher Faktor identifiziert wurde, der zur Bildung einer Temperaturschichtung unter der Flüssigkeitsoberfläche führt. Darüber hinaus wurde festgestellt, dass zur Minimierung der erforderlichen Druckgasmasse eine hohe Druckgastemperatur und/oder die Verwendung von Helium als Druckgas von Vorteil sind. Während der aktiven Bedrückung mit gasförmigem Stickstoff wurde Kondensation als vorherrschende Art des Phasenwechsels identifiziert und bei der Bedrückung mit Helium ist vornehmlich Verdampfung aufgetreten. Nach Ende der Bedrückung trat ausschließlich Kondensation auf. Für die Beschreibung dieses Druckabfalls wurde ein „smart fit" für die mit gasförmigem Stickstoff bedrückten Experimenten gefunden. Anhand dieser Gleichung wurde die Temperaturabnahme in der Gasphase als Haupttreiber des Druckabfalls identifiziert. Als dominierende Wärmeübertragung bei der Bedrückung wurde die Wärmeübertragung von dem Druckgas zur Tankwand bestätigt, die von der Druckgastemperatur abhängig ist. Die so zugeführte Wärme wird in der Tankwand in Richtung der flüssigen Phase geleitet und verstärkt den Temperaturanstieg in den wandnahen oberen Flüssigkeitsschichten. Unter der Berücksichtigung des Wärmeübergangs vom Druckgas zur Tankwand und bei Vernachlässigung des Phasenübergangs wurde eine Korrelation gefunden, die eine a priori Bestimmung des Druckanstieges im Tank für die durchgeführten Experimente ermöglicht. Des Weiteren wurde auf Basis der JAKOB Zahl und der FROUDE Zahl für thermische Ausdehnung eine Korrelation für die Bestimmung der benötigten Druckgasmasse gefunden. Die vorliegende Arbeit beschreibt im Detail die durchgeführten Bedrückungsexperimente, das numerische Modell sowie die analytische Ansätze und Korrelationen. Die Ergebnisse und die daraus gezogene Schlussfolgerungen werden vorgestellt und diskutiert.

Acknowledgement

Let all that I am praise the Lord;

with my whole heart, I will praise his holy name.

Let all that I am praise the Lord;

may I never forget the good things he does for me.

Psalm 103, 1-2

I would like to express my very great appreciation to Prof. Dr. Michael Dreyer, head of the Fluid Mechanics and Multiphase Flows group at ZARM, for his professional guidance, valuable support and constructive suggestions. His willingness to give his time so generously has been very much appreciated, especially since I was an external PhD student. I am particularly grateful that he provided me the opportunity to perform the experiments with the facility and the professional staff of ZARM.

In addition, I want to thank Prof. Dr. Andreas Rittweger, head of the DLR Institute of Space Systems in Bremen, that he agreed at short notice to be the second supervisor of this thesis. I am thankful for the very interesting conversations in which he shared his experience from industry on this field of work.

Moreover, I would like to offer my special thanks to Dr. Martin Sippel, head of the SART (Space Launcher Systems Analysis) group at the DLR Institute of Space Systems for giving me the opportunity to do my PhD in compatibility with my work at SART. I am thankful for a very interesting time and I greatly appreciate his support by providing me the computer and licenses for the Flow-3D simulations.

Additionally, I would like to thank all my colleagues from SART for the great time. I am particularly grateful for the support given by Carola Bauer as she has always offered her advice and Scott Fisher for the linguistic revision.

My special thanks are extended to Peter Prengel and Frank Ciecior for their effort in preparing and performing the experiments at ZARM. Moreover I would like to thank Diana Gaulke and Eckart Fuhrmann for providing me helpful advice for the numerical simulations. Additionally I am grateful to my former student Peter Friese for his excellent work and his assistance in performing the experiments.

Furthermore, I want to thank Prof. Dr. Emil Hopfinger from the Université Joseph Fourier in Grenoble for the very interesting and instructive time I was allowed to spend with him, his helpful advice as well as the great collaboration for the joint paper.

Finally, I would like to thank my parents, my sister and my whole family for their great support and prayer through all the times, as well as Felix Dorbath for all his advice, encouragement and love and that we could experience this time together.

Contents

List of Figures

List of Tables

Nomenclature

Roman letters

A	area, m^2
c_p	specific heat capacity at constant pressure, $J/(kg\ K)$
c_v	specific heat capacity at constant volume, $J/(kg\ K)$
D	diameter, m
D_t	thermal diffusivity, m^2/s
D_{ij}	diffusion coefficient (fluid i into fluid j), m^2/s
E	energy, J
E_{kin}	kinetic energy, J
E_{pot}	potential energy, J
e	energy per unit mass, J/kg
G	fluid property
g	gravitational acceleration, m/s^2
H	height, m
h	specific enthalpy, J/kg
Δh_v	specific enthalpy of vaporization, J/kg
J	characteristic mass flux, $kg/(m2\ s)$
j	vaporization rate, $kg/(m^2\ s)$
K	constant
L	characteristic linear dimension, m
l_n	standard liter, L
M	molecular mass, kg/mol

m	mass, kg
$\dot{m}$	mass flow rate, kg/s
$\hat{m}$	mass flux, kg/(m^2 s)
n	amount of substance, mol
p	pressure, Pa
p_{norm}	norm pressure $= 101.3$ kPa
p	pressure sensor
Q	heat, J
$\dot{Q}$	heat flow rate, W
$\dot{q}$	heat flux, W/m^2
R	tank radius, m
$\bar{R}$	universal gas constant, 8.3445 J/(mol K)
R_s	specific gas constant (GN2: 296.8 J/(kg K))
r	radial direction, m
s	thickness of a plate, m
T	temperature, K
T	temperature sensor
$\bar{T}$	mean temperature, K
$T_{pg,m}$	maximal pressurant gas inlet temperature (for this study: 352 K)
T_{ref}	reference temperature (LN2: 77.35 K)
t	time, s
U	internal energy, J
u	specific internal energy, J/kg
V	volume, m^3
V	valve
v	specific volume, m^3/kg
v	velocity, m/s
W	work, J
$\dot{W}$	rate of work, W
z	height coordinate, m

Greek letters

α	convection heat transfer coefficient, $W/(m^2\ K)$
β_Ψ	composition coefficient
β_T	thermal expansion coefficient, $1/K$
γ	intensive value or amount of G per unit mass
δ_T	thermal boundary layer thickness, m
Θ	temperature difference, K
ϵ	emissivity
λ	thermal conductivity, $W/(m\ K)$
μ	dynamic viscosity, Pa s
ρ	density, kg/m^3
$\rho_{ref,v}$	reference density (GN2: $4.61\ kg/m^3$)
σ	Stefan-Boltzmann constant, $5.67{\cdot}10^{-8}\ W/(m^2\ K^4)$
$\hat{\sigma}$	accommodation coefficient acc. to [18]
τ	characteristic time, s
ν	kinematic viscosity, m^2/s
ϕ_{GHe}	molar fraction of helium dissolved in LN2
χ	mole fraction
Ψ	mass fraction
$\dot{\omega}$	rate of energy generation per unit volume, W/m^3

Vectors and Tensors

$\boldsymbol{a}$	acceleration, m/s^2
$\boldsymbol{F}$	force, N
$\boldsymbol{N}$	normal unit vector of surface, m
$\boldsymbol{g}$	acceleration vector due to gravity, m/s^2
$\boldsymbol{G}$	fluid property
$\boldsymbol{n}$	normal unit vector
$\dot{\boldsymbol{q}}$	heat flux vector, W/m^2
$\boldsymbol{r}$	position vector

v	velocity, m/s
γ	intensive value
τ	viscous stress tensor, Pa
∇	Nabla operator

Subscripts

0	pressurization start
50	50 % ullage volume
b	bottom
bdy	body
cn	characteristic number
con	conduction
$cond$	condensation
$conv$	convection
CS	control surface
CV	control volume
dif	diffuser
$evap$	evaporation
exp	experimental
f	pressurization end
fl	fluid
$GH2$	gaseous hydrogen
GHe	gaseous helium
$GN2$	gaseous nitrogen
h	height
He	helium
i	initial
in	inflow
iso	isochoric
l	liquid
$LN2$	liquid nitrogen
m	maximal

$N2$	nitrogen
net	net
$norm$	norm
out	outflow
p	pressurization
pg	pressurant gas
phc	phase change
$press$	pressurization
r	relative
rad	radiation
ref	reference
$relax$	relaxation
$S1$, $S2$	side of plate
s	solid surface
sat	saturation
sh	shaft
sys	system
T	relaxation end
tot	total
u	ullage
v	shear
v	vapor
w	wall
Γ	phase interface
∞	bulk

Superscripts

$\sim$	scaling
$*$	nondimensional
sat	saturation

Abbreviations

CFD	computational fluid dynamics
CS	control surface
CV	control volume
DLR	Deutsches Zentrum für Luft- und Raumfahrt
ELDO	European Launcher Development Organization
EPC	Étage Principal Cryotechnique
EPS	Étage à Propergols Stockables
ESA	European Space Agency
ESC-A	Etage Supérieur Cryotechnique Type A
GH2	gaseous hydrogen
GHe	gaseous helium
GN2	gaseous nitrogen
GOX	gaseous oxygen
GSLV	Geosynchronous Satellite Launch Vehicle
HLV	Heavy Lift Launch Vehicle
KSLV	Korean Space Launch Vehicle
L3S	Lanceur 3ième Génération Substitution
LH2	liquid hydrogen
LN2	liquid nitrogen
LOX	liquid oxygen
ME	Mid-life Evolution
MMH	monomethylhydrazine
NASA	National Aeronautics and Space Administration
NASP	National Aero-Space Plane
NPSH	net positive suction head
NTO	nitrogen tetroxide
PMP	Propellant Management Program
SART	Space Launcher Systems Analysis
VOF	Volume of Fluid
ZARM	Center of applied space technology and microgravity

Nondimensional Numbers

$$\mathbf{Ev} = \frac{\Delta h_v \, J \, L}{\lambda_l \, \Theta}$$
Evaporative heat flux number

$$\mathbf{Ev_v} = \frac{\Delta h_v \, J \, L}{\lambda_v \, \Theta}$$
Evaporative heat flux number vapor phase

$$\mathbf{Fr}_{\beta_T} = \frac{v_{ref}^2}{g \, \beta_{T,l} \, \Theta \, L}$$
Thermal expansion FROUDE number

$$\mathbf{Fr} = \frac{v_{ref}^2}{g \, L}$$
FROUDE number

$$\mathbf{Fr}_{\beta_T\mathbf{v}} = \frac{v_{ref}^2}{g \, \beta_{T,v} \, \Theta \, L}$$
Thermal expansion FROUDE number vapor phase

$$\mathbf{Fr}_{\Psi} = \frac{v_{ref}^2}{g \, \beta_{\Psi} \, \Theta \, L}$$
Species expansion FROUDE number

$$\mathbf{Gr} = \frac{g \, \beta_T \, L^3 \, \Theta}{\nu^2}$$
GRASHOF number

$$\mathbf{Ja} = \frac{c_{p,l} \, \Theta}{\Delta h_v}$$
JAKOB number

$$\mathbf{Ja_v} = \frac{c_{p,v} \, \Theta}{\Delta h_v}$$
JAKOB number vapor phase

$$\mathbf{Nu} = \frac{\alpha L}{\lambda_l}$$
NUSSELT number

$$\mathbf{Nu_v} = \frac{\alpha L}{\lambda_v}$$
NUSSELT number vapor phase

$$\mathbf{Pr} = \frac{\mu_l \, c_{p,l}}{\lambda_l}$$
PRANDTL number

$$\mathbf{Pr_v} = \frac{\mu_v \, c_{p,v}}{\lambda_v}$$
PRANDTL number vapor phase

$$\mathbf{Ra} = \mathbf{Gr} \, \mathbf{Pr} = \frac{g \, \beta_T \, L^3 \, \Theta}{\nu \, D_t}$$
RAYLEIGH number

$$\mathbf{Re} = \frac{\rho_{l,ref} \, v_{ref} \, L}{\mu_l}$$
REYNOLDS number

$$\mathbf{R} = \frac{\rho_l}{\rho_v}$$
Density ratio

$$\mathbf{Re_v} = \frac{\rho_{v,ref} \, v_{v,ref} \, L}{\mu_v}$$
REYNOLDS number vapor phase

$$\mathbf{Sc_v} = \frac{\mu_v}{\rho_{v,ref} \, D_{i,j}}$$
SCHMIDT number vapor phase

$$\mathbf{St} = \frac{L}{\tau \, v_{ref}}$$
STROUHAL number

Chapter 1
INTRODUCTION

Today, liquid propellant rocket engines are commonly applied in the space segment all over the world. The idea of liquid propellant rockets was conceived over 100 years ago by pioneers as Konstantin Ziolkowski and Hermann Oberth. The American professor, physicist and inventor Robert H. Goddard is credited with developing and building the world's first liquid propellant rocket. He started testing the first liquid-fueled rocket engine in 1923 and after further developments and experiments, he launched the first liquid propellant rocket with gasoline and liquid oxygen as propellants in 1926 in Auburn, Massachusetts. In Europe, Johannes Winkler successfully launched the first liquid-fueled rocket in 1931 on the parade-ground near Dessau-Großkühnau in Germany, using the propellant combination liquid oxygen and liquid methane. In 1942, the first large, flight controlled and stabilized rocket Aggregat 4, later called V2, lifted off in Peenemünde, Germany, using an alcohol-water mixture and liquid oxygen as propellants [94, 98].

The technology of liquid propellant rocket engines is much more refined today. Various propellants were developed and tested and the liquid propellant combination with the highest specific impulse is liquid oxygen (LOX) and liquid hydrogen (LH2). It is the most famous cryogenic propellant combination for launcher applications and focus of this study. LOX was the earliest, cheapest, safest and eventually the preferred oxidizer for large launch vehicles. Its main drawback is that it is moderately cryogenic, which means that it has to be stored at very low temperatures in order to keep it liquefied. LH2 was identified by all the leading rocket visionaries as the theoretically ideal rocket fuel. However, its big drawbacks are that it is highly cryogenic and has a very low density, requiring large tanks.

In the mid-1950s, the USA mastered hydrogen technology and developed the word's first cryogenic launcher upper stage, the Centaur. Relevant cryogenic launcher stages which were consequently developed in the USA were the second and third stage of the Saturn V launcher, which were using LOX and LH2 as propellant combination and enabled a manned landing on the Moon. Furthermore, the NASA Space Transportation System, better known as the Space Shuttle, used LOX and LH2, stored in the external tank, for the operation of the orbiter main engines. For the first three flights of the Indian launcher GSLV, the 12KRB cryogenic third stage from a Russian manufacturer was used. This stage is based on the KVRB upper stage, which is currently under development for the Russian launchers Proton and Angara. In 2010, the first flight test of the indigenous designed and built Indian cryogenic upper stage failed, however in January 2014 it was launched successfully. Further cryogenic stages, which already came into operation are the first stage of the Russian Energiya launcher, the first and second stage of the Japanese H-II launcher and the third stage of the Chinese launcher Long March 3 (also called CZ-3). For the new generation of the Long March rocket family Long March 5 to 7 a cryogenic upper stage is currently under development.

The first development of new technologies for a cryogenic launcher stage in Europe began in 1967 under the name of LH2 Experimental Program, coordinated by the European Launcher Development Organization (ELDO). It was experimental work as preparation for the development of the new Europa 3 launcher with its LOX/LH2 second stage. However in 1972, the European Space Conference decided to cease of work on the Europa 3 launcher in favor of the French proposal L3S (Lanceur 3ième Génération Substitution). This launcher concept envisaged a LOX/LH2 third stage, called H6, which was much smaller than the Europa 3 upper stage concept. In 1973, the L3S concept was improved and the launcher name was changed to Ariane [102]. Six years later, on the 24th of December 1979, the first European launcher Ariane 1 lifted off the spaceport in Kourou in French Guiana. The Ari-

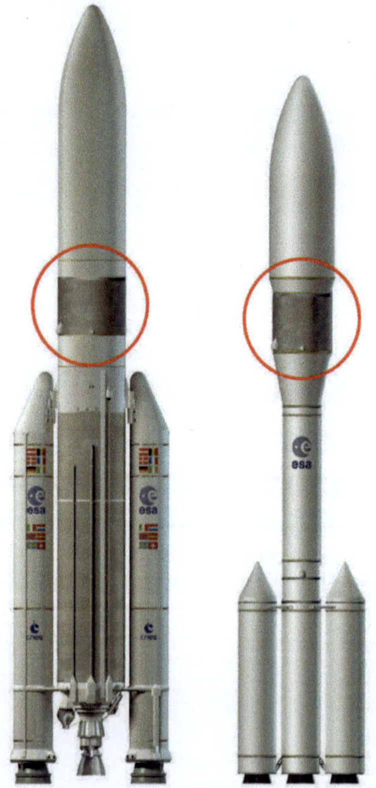

Fig. 1.1 Proposal for an Adapted Ariane 5 ME (left) and proposal for Ariane 6 (right) [41]. The red circles indicate the cryogenic upper stages.

ane 1, used the first HM-7 engine using LOX and LH2. This upper stage concept was applied with some modifications for the Ariane 2, Ariane 3 and Ariane 4 launchers [61]. The current Ariane 5 launcher can use either the cryogenic upper stage ESC-A (Etage Supérieur Cryotechnique Type A) with LOX and LH2 as propellants, or the re-ignitable upper stage EPS (Étage à Propergols Stockables) with the storable propellants NTO and MMH. The main stage EPC (Étage Principal Cryotechnique) of the Ariane 5 launcher also uses the cryogenic propellants LOX and LH2.

Since 2008 ESA member states have been developing an improved type of the current launcher Ariane 5, named Adapted Ariane 5 ME (Mid-life Evolution), schematically depicted in Figure 1.1. This launcher is intended to have an increased payload capacity which should be enabled, amongst others, by a new cryogenic upper stage. This stage should be propelled with approximately 28 tons of LOX and LH2 and use the Vinci engine, which is currently under development. In November 2012, the ESA ministerial council committed to the continuation of the Adapted Ariane 5 ME program for another two years. They also initiated detailed definition studies of another new launcher concept, called Ariane 6 (schematically depicted in Figure 1.1). Further decision on the future of both launcher concepts shall be taken at the end of 2014. Today's baseline is that both, the Adapted Ariane 5 ME and the Ariane 6 concept, shall each include a newly developed cryogenic upper stage, based on the restartable cryogenic engine Vinci (positions marked in Figure 1.1 by the red circles) [87]. This baseline is significant for Germany and especially for the Bremen site, which is a prominent player in upper stage research and development.

Against the background of the new development of European cryogenic upper stages for the new launcher generation, interest in the advancement of cryogenic fluid management technologies has been significantly revived. Included in cryogenic fluid management technologies is the the propellant feed system, the task of which is to feed the liquid propellant from the propellant tank to the rocket engine thrust chamber at pre-defined mass flow rates and required conditions, such as temperature and pressure. Two distinct types of propellant feed systems can be defined: the pressure-fed and the pump-fed systems. For a pressure-fed propellant feed systems, the engine thrust is directly proportional to the tank ullage pressure. Therefore, relatively high tank pressures of $1.3 \cdot 10^3$ kPa up to $9 \cdot 10^3$ kPa are required, as the propellant tank pressure needs to be higher than the pressure in the engine combustion chamber. In a pump-fed propellant feed system, lower tank pressures

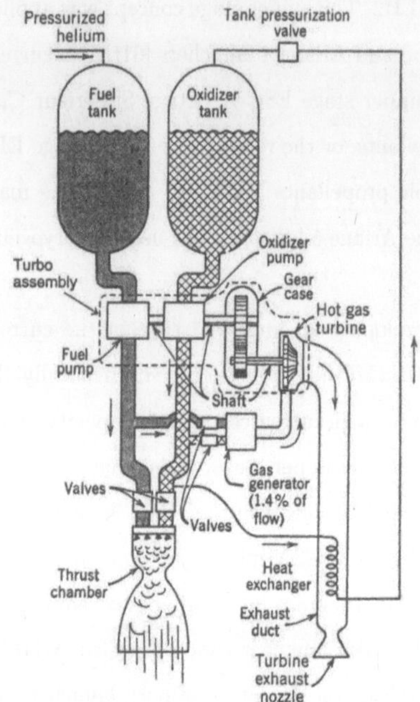

Fig. 1.2 Flow diagram of a liquid propellant rocket engine with stored-inert-gas and evaporated-propellant pressurization systems [94].

(between $0.07 \cdot 10^3$ kPa and $0.34 \cdot 10^3$ kPa) are used, as turbopumps are applied to raise the propellant pressure to a level above tank pressure, which is suitable for the injection into the engine combustion chamber [95].

One component of the pump-fed propellant feed system is the tank pressurization system, which forms the focus of this study. The purpose of the pressurization system is to control and maintain the required pressure in the propellant tanks at any time of the mission. The defined tank pressure history is bounded by propellant and tank structural requirements, the thrust profile and the net positive suction head (NPSH). The NPSH is the minimum turbopump suction head which is required to avoid cavitation throughout the operating cycles. Commonly, three different tank pressurization systems are distinguished: the combustion-products, the evaporated-propellant and the stored-inert-gas system.

For the combustion-products system, the pressurant gas is obtained by combustion of propellants. E.g., a gas generator produces fuel-rich exhaust gas for the operation of the engine turbopump and at the outlet of the turbine, gas for the tank pressurization is tapped off, cooled down and fed back into the propellant tank. This system has been used only for few applications, as the pressurant gas is often chemically incompatible with the propellant, has a too high temperature or condensable elements.

The evaporated-propellant pressurization system appears in two versions: the simplest is the self-pressurization by the propellant's own boil-off vapor. This system in general requires a propellant with a high vapor pressure and the resulting tank pressure is very dependant on the mission profile. The system is very reliable, but requires high pressurant gas masses due to low pressurant gas temperatures and therefore high fluid densities. The second and most common form of the evaporated-propellant pressurization system is active-pressurization with evaporated propellant. A part of the liquid propellant is tapped off at the pump discharge area and routed through a heat exchanger. Within this heat exchanger, the propellant is vaporized and then led as pressurant gas back into its own propellant tank [72]. Such system is schematically depicted in Figure 1.2 for the oxidizer propellant tank.

The third, widely used system for pressurizing a propellant tank is the stored-inert-gas system, as schematically depicted in Figure 1.2 for the fuel tank, where helium is applied as a pressurant gas. For the stored-inert-gas system, the applied pressurant gas is stored in external vessels, preferably at low temperature, resulting in high storage density. Pressurizing with an inert gas results in the fact, that during and after pressurization, the tank ullage includes both, the inert pressurant gas and a quantity of evaporated propellant.

This study focuses on both, active-pressurization with an evaporated-propellant system and a stored-inert-gas pressurization system, as they are intended to be applied for the future European cryogenic upper stage concepts.

Chapter 2
THEORETICAL BACKGROUND

This chapter provides an overview of the theoretical background of this study, such as the relevant fundamental equations of fluid dynamic and thermodynamics, heat transfer and phase change mechanisms as well as thermal stratification. It also includes the derivation of an analytical equation for the pressure drop and the scaling concept with relevant characteristic numbers.

Figure 2.1 displays a typical propellant tank, which is cylindrical with an ellipsoidal shaped bottom. It is partly filled with liquid propellant and pressurant gas is injected through an inlet in the tank lid. The z-axis points in the opposite direction to the gravity vector $\boldsymbol{g} = (0,0,-g)^T$ and the origin of the coordinate system is set to the bottom of the tank. Liquid and ullage is separated by a free surface and the normal vector, perpendicular to it, is set as $\boldsymbol{n}$, pointing in the positive z direction.

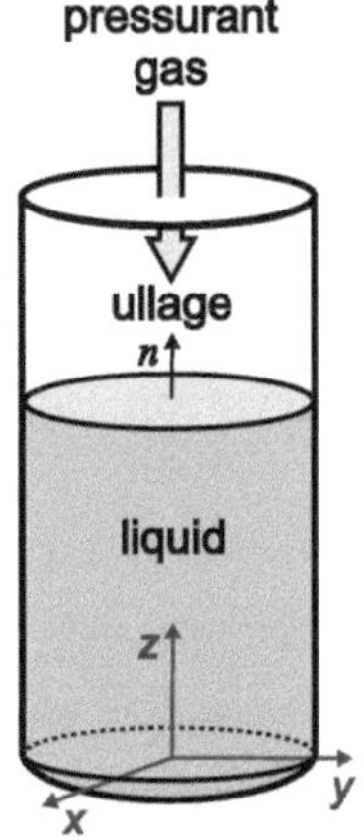

Fig. 2.1 Schematic of the considered cylindrical propellant tank with ellipsoidal shaped bottom, partly filled with liquid propellant. The pressurant gas is injected through an inlet in the lid into the tank ullage.

2.1 Ideal Gas

The thermodynamic properties of an ideal gas can be described by the ideal gas law

$$p\,V = n\,\bar{R}\,T \tag{2.1}$$

with p as the pressure, V the volume, n is the amount of substance of the gas, $\bar{R}$ the universal gas constant and T is the temperature.

In a mixture of ideal gases, each gas has a partial pressure. The partial pressure p_i of one component is determined as

$$p_i = p \, \chi_i \qquad (2.2)$$

with χ_i as the mole fraction of that gas component. The mole fraction is defined as

$$\chi_i = \frac{n_i}{n} \qquad (2.3)$$

where n_i is the amount of substance of the gas component i and n is the total amount of substance. The mass fraction of an ideal gas mixture is defined as

$$\Psi_i = \frac{m_i}{m} \qquad (2.4)$$

where m_i is the mass of one gas component and m the total gas mass.

2.2 Conservation Equations

In order to describe a typical fluid dynamics problem, the mode of change over time of the system's physical variables needs to be determined. This is achieved by the Reynolds' transport theorem from which the basic conservation equations, the conservation of mass, the conservation of momentum and the conservation of energy are derived. In the following sections, these conservation equations are introduced together with the Reynolds' transport theorem, based on White [103] and Bird [13]. In accordance with the conventional definition, the presented governing equations describe a one-phase system with liquid, which is considered as a continuum with a free surface.

2.2.1 Reynolds' Transport Theorem

The Reynolds' transport theorem is used to characterize the dynamics of a physical variable and its change over time. It states that the total change over time of any fluid property G in a volume V that moves with the flow is equal to the change over time in the current control volume CV plus the flux of G passing through the instantaneous surface area A of the control volume. For $\boldsymbol{n}$ being the outward normal unit vector on the control surface, the Reynolds' transport theorem can be written in the compact form as

$$\frac{d}{dt}(G_{sys}) = \frac{d}{dt}\left(\int_{CV} \gamma \rho dV \right) + \int_{CS} \gamma \rho (\boldsymbol{v} \bullet \boldsymbol{n}) dA \qquad (2.5)$$

with CV being an arbitrary fixed control volume, CS the control surface, t the time, ρ the density $\gamma = dG/dm$ the intensive value (i.e. the amount of the arbitrary mass-specific parameter G in any small portion of the fluid) and with $\boldsymbol{v}$ being the fluid velocity.

2.2.2 Conservation of Mass

The equation for the conservation of mass can be derived from the Reynolds' transport theorem (Equation 2.5). This conservation equation states that the mass of a closed system m_{sys} remains constant over time. For the integral form of the equation for conservation of mass, the variable G becomes the mass m. It follows $\gamma = dm/dm = 1$. The equation for a deformable control volume is therefore

$$\left(\frac{dm}{dt}\right)_{sys} = 0 = \frac{d}{dt}\left(\int_{CV} \rho dV\right) + \int_{CS} \rho(\boldsymbol{v}_r \bullet \boldsymbol{n})dA \tag{2.6}$$

where the relative velocity $\boldsymbol{v}_r$ is defined as $\boldsymbol{v}_r = \boldsymbol{v}(\boldsymbol{r},t) - \boldsymbol{v}_{CV}(t)$ if the control volume is moving uniformly at velocity $\boldsymbol{v}_{CV}(t)$. The vector $\boldsymbol{r}$ is the position vector and t is the time. As previously stated, CS is the control surface of the mass flux. For a fixed control volume, it follows that

$$\int_{CV} \frac{\partial \rho}{\partial t}dV + \int_{CS} \rho(\boldsymbol{v} \bullet \boldsymbol{n})dA = 0. \tag{2.7}$$

If the control volume only has an i number of one-dimensional inlets and outlets, it can be written as:

$$\int_{CV} \frac{\partial \rho}{\partial t}dV + \sum_i (\rho_i A_i v_i)_{out} - \sum_i (\rho_i A_i v_i)_{in} = 0 \tag{2.8}$$

For incompressible flows, where $\frac{\partial \rho}{\partial t} \approx 0$, Equation 2.7 becomes

$$\int_{CS} \rho(\boldsymbol{v} \bullet \boldsymbol{n})dA = 0. \tag{2.9}$$

The differential form of mass conservation for an infinitesimal fixed control volume (dx, dy, dz) follows from Equation 2.7. The considered volume element is so small that the volume integral reduces to a differential term.

$$\int_{CV} \frac{\partial \rho}{\partial t}dV \approx \frac{\partial \rho}{\partial t} dx\, dy\, dz \tag{2.10}$$

Equation 2.10 is the equation of mass conservation, which describes the rate of change of density at a fixed point resulting from the changes in the mass velocity vector $\rho\boldsymbol{v}$. It follows that the compact form for the differential equation of mass conservation, written in the vector symbolism, is

$$\frac{\partial \rho}{\partial t} + \nabla \cdot (\rho \boldsymbol{v}) = 0 \tag{2.11}$$

where $\nabla \cdot (\rho \boldsymbol{v})$ is the divergence of the mass flux $\rho \boldsymbol{v}$. For incompressible flows this results in

$$\nabla \cdot \boldsymbol{v} = 0. \tag{2.12}$$

2.2.3 Conservation of Momentum

The law of conservation of momentum states that the total momentum of a system remains unchanged if the system is closed. If the surroundings exert a net force $\boldsymbol{F}$ on the system, Newton's second law states that its mass will begin to accelerate.

$$\boldsymbol{F} = m\boldsymbol{a} = m\frac{d\boldsymbol{v}}{dt} = \frac{d}{dt}(m\boldsymbol{v}) \tag{2.13}$$

Consequently, it follows from the Reynolds' transport theorem (Equation 2.5), with $\boldsymbol{G} = m\boldsymbol{v}$ being the linear momentum and $\boldsymbol{\gamma} = d\boldsymbol{G}/dm = \boldsymbol{v}$

$$\frac{d}{dt}(m\boldsymbol{v})_{sys} = \sum \boldsymbol{F} = \frac{d}{dt}\left(\int_{CV} \boldsymbol{v}\rho dV\right) + \int_{CS} \boldsymbol{v}\rho(\boldsymbol{v}_r \cdot \boldsymbol{n})dA \tag{2.14}$$

where the term $\boldsymbol{v}$ is the fluid velocity relative to an inertial coordinate system and the term $\sum \boldsymbol{F}$ is the vector sum of all forces acting on the control volume. If the control volume only has an i number of one-dimensional inlets and outlets, Equation 2.14 reduces to the following form

$$\sum \boldsymbol{F} = \frac{d}{dt}\left(\int_{CV} \boldsymbol{v}\rho dV\right) + \sum(\dot{m}_i\boldsymbol{v}_i)_{out} - \sum(\dot{m}_i\boldsymbol{v}_i)_{in} \tag{2.15}$$

with $\dot{m}_i$ as the mass flow rates.

The differential form of the equation of linear momentum is derived from Equation 2.15 with the assumption that one element is so small that the volume integral reduces to a derivative term.

$$\frac{\partial}{\partial t}\left(\int_{CV} \boldsymbol{v}\rho dV\right) \approx \frac{\partial}{\partial t}(\rho \boldsymbol{v}) \, dx \, dy \, dz \tag{2.16}$$

By considering gravity $\boldsymbol{g}$ as the only body force and $\boldsymbol{\tau}_{ij}$ as the viscous-stress tensor acting on an element, the basic differential momentum equation for an infinitesimal element follows

$$\rho\left(\frac{\partial \boldsymbol{v}}{\partial t} + (\boldsymbol{v} \cdot \nabla)\boldsymbol{v}\right) = -\nabla p - (\nabla \cdot \boldsymbol{\tau}) + \rho \boldsymbol{g} \tag{2.17}$$

where ∇p is the pressure force, $(\nabla \cdot \boldsymbol{\tau})$ the viscous force and $\rho\boldsymbol{g}$ is the gravitational force. Each of these are defined on an infinitesimal element per unit volume. For an incompressible flow with constant density ρ and constant viscosity μ, the Navier-Stokes equation follows from Equation 2.17.

$$\rho\left(\frac{\partial \boldsymbol{v}}{\partial t} + (\boldsymbol{v} \cdot \nabla)\boldsymbol{v}\right) = -\nabla p + \mu\nabla^2\boldsymbol{v} + \rho\boldsymbol{g} \qquad (2.18)$$

For nonviscous flows, the Euler equation follows from Equation 2.17.

$$\rho\left(\frac{\partial \boldsymbol{v}}{\partial t} + (\boldsymbol{v} \cdot \nabla)\boldsymbol{v}\right) = -\nabla p + \rho\boldsymbol{g} \qquad (2.19)$$

2.2.4 Conservation of Energy

The equation for the conservation of energy states that a system's energy changes only by adding or dissipating energy over the system boundary. The first law of thermodynamics states that if heat dQ is added to the system or work dW is done by the system, the system energy dE must change:

$$\frac{dQ}{dt} - \frac{dW}{dt} = \frac{dE}{dt} \qquad (2.20)$$

This is applied to the Reynolds' transport theorem (Equation 2.5) with the variable G equal to E and the energy per unit mass $\gamma = dE/dm = e$. For a fixed control volume, Equation 2.20 can be written as

$$\frac{dQ}{dt} - \frac{dW}{dt} = \frac{dE}{dt} = \frac{d}{dt}\left(\int_{CV} e\rho dV\right) + \int_{CS} e\rho(\boldsymbol{v} \cdot \boldsymbol{n})dA. \qquad (2.21)$$

A positive Q means that heat is added to the system and a positive W denotes, that the work is done by the system. The energy per unit mass e consists in most cases of an internal, a kinetic and a potential part.

$$e = u + \frac{1}{2}v^2 + gz \qquad (2.22)$$

Therefore, Equation 2.21 takes the following form.

$$\dot{Q} - \dot{W}_{sh} - \dot{W}_v = \frac{\partial}{\partial t}\left[\int_{CV}\left(u + \frac{1}{2}v^2 + gz\right)\rho dV\right] + \int_{CS}\left(h + \frac{1}{2}v^2 + gz\right)\rho(\boldsymbol{v} \cdot \boldsymbol{n})dA \quad (2.23)$$

In this equation $\dot{W}_{sh}$ stands for the rate of shaft work, $\dot{W}_v$ is the rate of shear work due to viscous stresses and the pressure work is included in the last term on the right side. The parameter u is the specific internal energy of the system and h is the specific enthalpy. For a steady flow having a number of inlets and outlets, which are assumed as one-dimensional, it follows

$$\int_{CS} \left(h + \frac{1}{2}v^2 + gz \right) \rho(\boldsymbol{v} \bullet \boldsymbol{n})dA = \sum (h + \frac{1}{2}v^2 + gz)_{out}\, \dot{m}_{out} - \sum (h + \frac{1}{2}v^2 + gz)_{in}\, \dot{m}_{in} \quad (2.24)$$

where the values of h, $\frac{1}{2}v^2$ and gz are assumed to be averages over each cross section. The specific enthalpy h of the fluid, which is always dependent on a reference state, is defined as

$$h := u + pv \qquad (2.25)$$

with v as the specific volume.

Based on Equation 2.23, the differential equation of energy for an infinitesimal fixed control volume with $\dot{W}_{sh} = 0$, incompressible fluids, negligence of the dissipation term and by considering only the heat conduction and using Fourier's law (see Equation 2.38) follows

$$\rho c_v \left(\frac{\partial T}{\partial t} + (\boldsymbol{v} \bullet \nabla)T \right) = \lambda \nabla^2 T \qquad (2.26)$$

with λ as the thermal conductivity and c_v as the specific heat capacity at constant volume.

2.2.5 First Law of Thermodynamics

According to the law of conservation of energy, a system's energy changes only by adding or dissipating energy over the system boundary.

2.2.5.1 Closed System

For a closed system, the energy of the system must remain constant. The system's total energy E_{sys} consists of the internal energy U, the kinetic energy E_{kin} and the potential energy E_{pot}. The total energy balances the thermal energy Q and the work W. Heat added to and work done by the system are denoted with a positive sign.

$$dE_{sys} = dU + dE_{kin} + dE_{pot} = dQ + dW \qquad (2.27)$$

For a stationary closed system, kinetic and potential energy are not changing. It therefore follows that the energy balance equation for a stationary closed system is:

$$dU = dQ + dW. \qquad (2.28)$$

The parameter W contains work, which results in a change of the internal state of the system such as the work for change of volume, electrical work or shaft work.

The internal energy of an ideal gas, which is always dependent on a reference state, is defined as

$$U = mc_v T \tag{2.29}$$

with c_v as the specific heat capacity at constant volume.

In this study, results of active-pressurization experiments are presented. In these experiments, the fluid temperatures are measured using multiple temperature sensors. For the analysis of the experimental results in Section 6.5.1, the mean temperature and mass of the vapor phase has to be determined. It is assumed that the temperature distribution between two sequencing temperature sensors in the experiments is linear. Therefore, the following equation can be used for the mean internal energy of the vapor phase U_v between two temperature sensors.

$$U_v = c_v \int_{z_1}^{z_2} \rho_v(z)\, T_v(z)\, A\, dz \tag{2.30}$$

The parameter z is the variable for the height with $z = 0$ at the tank bottom. In Equation 2.30, z_1 is the distance from the tank bottom to one vapor temperature sensor and z_2 is the distance from the tank bottom to the superior sensor. With this approach follows for the average vapor temperature $\bar{T}_v$ of the vapor phase

$$\bar{T}_v = \frac{c_v \pi R^2 \left(\int_{z_1}^{z_2} T_v(z)\rho_v(z)dz + \int_{z_2}^{z_3} T_v(z)\rho_v(z)dz + \dots \right)}{c_v\, m_v} \tag{2.31}$$

where $T_v(z)$ is the linear temperature distribution between two sequencing temperature sensors, R is the inner radius of the tank and z_1 to z_n are the positions of the temperature sensors, placed in the vapor phase.

The vapor mass m_v can then be calculated based on the equation for the density $m = \rho\, V$:

$$m_v = \pi R^2 \left(\int_{z_1}^{z_2} \rho_v(z)dz + \int_{z_2}^{z_3} \rho_v(z)dz + \dots \right) \tag{2.32}$$

2.2.5.2 Open System

For any unsteady process of an open system, the energy balance can be written as follows:

$$dE_{sys} = dU + dE_{kin} + dE_{pot} = dQ + dW + \sum m_{in} \left(h + \frac{v^2}{2} + gz \right)_{in}$$
$$- \sum m_{out} \left(h + \frac{v^2}{2} + gz \right)_{out} \tag{2.33}$$

The parameter m_{in} and m_{out} are the masses entering respectively leaving the control volume with the specific enthalpy h, the kinetic energy $\frac{v^2}{2}$ and the potential energy gz. The derivative with respect to time follows then as

$$\frac{dE_{sys}}{dt} = \sum_i \dot{Q}_i + \sum_j \dot{W}_j + \sum \dot{m}_{in}\left(h + \frac{v^2}{2} + gz\right)_{in} - \sum \dot{m}_{out}\left(h + \frac{v^2}{2} + gz\right)_{out} \quad (2.34)$$

where $\dot{m}_{in}$ are the mass flow rates entering the system and $\dot{m}_{out}$ are those leaving. $\dot{Q}_i$ is the heat flow rate over the system boundary and $\dot{W}_j$ the work rate over the system boundary.

For the analysis of the performed experiments, the energy balance is applied: The total heat, entering the test tank Q can be split up into the heat entering the liquid phase Q_l and the heat going into the vapor phase Q_v.

$$dQ = dQ_l + dQ_v \quad (2.35)$$

Figure 6.13 depicts the considered tank system with the two applied separate control volumes for the vapor phase CV_v and the liquid phase CV_l. Based on the first law of thermodynamics for an open system, the energy balance for the liquid control volume can be written as

$$dQ_l = dU_l - \underbrace{\sum m_{l,in}\, h_{l,in}}_{condensation} + \underbrace{\sum m_{l,out}\, h_{l,out}}_{evaporation} \quad (2.36)$$

where the work, kinetic and potential energy terms have been disregarded.

For the vapor control volume the first law of thermodynamics for an open system has the following form for the pressurization phase.

$$dQ_v = dU_v - \underbrace{\sum m_{v,in}\, h_{v,in}}_{evaporation\,+\,press.gas} + \underbrace{\sum m_{v,out}\, h_{v,out}}_{condensation} \quad (2.37)$$

For the relaxation phase where no more pressurant gas is injected, the term for mass added by the pressurant gas is omitted.

2.3 Heat Transfer

Heat is always transferred from a higher-temperature medium to the lower-temperature medium until both are in equilibrium. In general, three different mechanisms of heat transfer can be distinguished: conduction, convection and radiation. All of these mechanisms are briefly described in this section. The presented equations are based on Polifke [77], Cengel [19] and White [103].

2.3.1 Conduction

Heat conduction represents a heat flow in a solid or fluid due to a temperature difference. The most important equation is the Fourier's law, which can be written as

$$\dot{q}_{con} = -\lambda \nabla T \tag{2.38}$$

where λ is the thermal conductivity which relates the vector of the heat flux $\dot{q}$ to the vector gradient of temperature ∇T.

Based on the conservation of energy and Fourier's law, the Fourier differential equation follows for the liquid phase as

$$\frac{\partial}{\partial x}\left(\lambda\frac{\partial T}{\partial x}\right) + \frac{\partial}{\partial y}\left(\lambda\frac{\partial T}{\partial y}\right) + \frac{\partial}{\partial z}\left(\lambda\frac{\partial T}{\partial z}\right) + \dot{\omega} = \rho\, c_p\, \frac{\partial T}{\partial t} \tag{2.39}$$

where $\dot{\omega}$ is the rate of energy generation per unit volume, ρ is the density and c_p is the specific heat capacity.

A material property which commonly appears in heat conduction analysis is the thermal diffusivity D_t

$$D_t = \frac{\lambda}{\rho\, c_p} \tag{2.40}$$

which is the thermal conductivity λ divided by the density ρ and the specific heat capacity c_p.

For a one-dimensional, nonsteady temperature field without heat sources, Equation 2.39 takes the following form.

$$\frac{\partial T}{\partial t} = D_t\, \frac{\partial^2 T}{\partial x^2} \tag{2.41}$$

This equation can be solved analytically for large media: suppose a semi-infinite block, initially at temperature T_0. For the time $t > 0$, the temperature at the surface of the block $x = 0$ is suddenly increased to T_1. This leads to the following equation

$$T - T_0 = (T_1 - T_0)\,\mathrm{erfc}\left[\frac{x}{2\sqrt{D_t\, t}}\right] \tag{2.42}$$

with erfc as the complementary error function, defined as $\mathrm{erfc}(x) = 1 - \mathrm{erf}(x)$.

For steady-state heat conduction in a flat plate, without consideration of heat sources, Equation 2.39 takes the following form.

$$\frac{\partial^2 T}{\partial x^2} = 0 \tag{2.43}$$

For a flat plate with the thickness s and the surface temperatures T_{S1} on the one side ($x = 0$) and T_{S2} on the other side of the plate ($x = s$), Equation 2.43 has the following solution.

$$T(x) = T_{S1} + (T_{S2} - T_{S1})\,\frac{x}{s} \qquad (2.44)$$

2.3.2 Convection

Heat transfer by convection represents an energy transfer between a solid surface and an adjacent liquid or gas in motion. It includes the combined effects of heat conduction and fluid motion. The convection heat transfer increases with ascending fluid motion. But if the fluid bulk is quiescent, pure conduction remains as a heat transfer mechanism between the solid surface and the adjacent fluid.

On one hand, there is forced convection, which occurs when the fluid is forced to flow over the surface by external means. On the other hand, there is natural or free convection, which appears if the fluid motion is caused by buoyancy forces, induced by density differences due to the temperature variations in the fluid. Heat transfer processes that include a phase change of a fluid are also considered as convection because of a fluid motion generated during this process. The convective heat flux can be written with Newton's law of cooling.

$$\dot{q}_{conv} = \alpha\,(T_s - T_\infty) \qquad (2.45)$$

Here, α is the convection heat transfer coefficient, T_s is the temperature of the solid surface and T_∞ is the temperature of the fluid sufficiently far away from the surface.

In order to determine the mode of convection, empirical correlations for the average NUSSELT number $\mathbf{Nu}$ can be applied. For the natural convection over surfaces follows according to Cengel [19] for a RAYLEIGH number of $10^4 < \mathbf{Ra} < 10^9$ (laminar)

$$\mathbf{Nu} = 0.59\,\mathbf{Ra}^{1/4} \qquad (2.46)$$

and for a RAYLEIGH number of $10^9 < \mathbf{Ra} < 10^{13}$ (turbulent)

$$\mathbf{Nu} = 0.1\,\mathbf{Ra}^{1/3}. \qquad (2.47)$$

A vertical cylinder can be considered as a flat plate if

$$D \geq \frac{35\,H}{\mathbf{Gr}^{1/4}} \qquad (2.48)$$

with D as the diameter and H as the height of the cylinder and $\mathbf{Gr}$ as the GRASHOF number of the fluid.

For a forced convection flow follows according to Cengel [19] for a REYNOLDS number of $\mathbf{Re} < 5 \cdot 10^5$ (laminar)

$$\mathbf{Nu} = 0.664\ \mathbf{Re}^{1/2}\ \mathbf{Pr}^{1/3} \tag{2.49}$$

and for $0.6 \le \mathbf{Pr} \le 60$ and $5 \cdot 10^5 \le \mathbf{Re} \le 10^7$ (turbulent)

$$\mathbf{Nu} = 0.037\ \mathbf{Re}^{0.8}\ \mathbf{Pr}^{1/3} \tag{2.50}$$

with $\mathbf{Pr}$ as the PRANDTL number.

2.3.3 Radiation

Heat transfer from a body due to emission of electromagnetic waves is known as thermal radiation. It is a consequence of the thermal agitation of the body's composing molecules. The heat flux caused by thermal radiation of a gray body with the surface temperature T_s is defined by the Stefan–Boltzmann law as

$$\dot{q}_{rad} = \epsilon\ \sigma\ T_s^{\,4} \tag{2.51}$$

where ϵ is the emissivity of the surface ($\epsilon = 1$ for a black body) and σ is the Stefan-Boltzmann constant $\sigma = 5.67 \cdot 10^{-8}\ \dfrac{\text{W}}{\text{m}^2\text{K}^4}$.

2.4 Phase Change

A phase change describes a phase transition from one phase or state of matter to another. This study only considers the transition between the liquid and the vapor phases. The temperature, at which a fluid can undergo phase change is the saturation temperature, corresponding to the saturation pressure. If a liquid's temperature increases over the saturation temperature, evaporation or boiling occurs. If the temperature of a gas falls below the saturation temperature, condensation occurs. The relationship between the saturation temperature and saturation pressure is described by the Clausius-Clapeyron equation.

$$\ln\left(\frac{p_1}{p_2}\right) = \frac{\Delta h_v}{R_s}\left(\frac{1}{T_2} - \frac{1}{T_1}\right) \tag{2.52}$$

In this equation, the subscripts 1 and 2 correspond to different locations on the saturation curve and R_s is the mass specific gas constant. A constant specific enthalpy of evaporation Δh_v in the considered pressure range is assumed.

In order to evaluate the mass of the fluid that undergoes phase change on the liquid-vapor interface, the free surface is considered at molecular level. Based on the kinetic theory of gases and with the Maxwell velocity distribution, which describes the statistical distribution of the amount of particle velocities in an ideal gas, the analysis of the interfacial transport assumes that the fluxes of vaporizing and condensing molecules can be derived separately for each flux direction. The superimposed results obtain the net flux. Although this is an approximative approach, its results provide an estimate of the effects of the interfacial molecular transport on vaporization and condensation processes. Based on Carey [18], the interfacial mass flux $\hat{m}_{phc}$ can be determined by the following equation

$$\hat{m}_{phc} = \left[\frac{2\hat{\sigma}}{(2-\hat{\sigma})} \right] \sqrt{\frac{M}{2\pi\bar{R}}} \left(\frac{p_v}{\sqrt{T_v}} - \frac{p_l}{\sqrt{T_l}} \right) \tag{2.53}$$

where M is the molar mass, $\bar{R}$ is the universal gas constant, p_v and T_v are the vapor pressure and temperature and p_l and T_l the pressure and temperature at the liquid surface. If the vapor and liquid temperature at the fluid interface are considered as equal, Equation 2.53 can be simplified as follows.

$$\hat{m}_{phc} = \left[\frac{2\hat{\sigma}}{(2-\hat{\sigma})} \right] \sqrt{\frac{1}{2\pi R_s}} \frac{p_v - p_{l,sat}}{\sqrt{T_\Gamma}} \tag{2.54}$$

Where T_Γ is the temperature at the liquid surface, $p_{l,sat}$ is the related saturation pressure of the liquid and R_s the specific gas constant. The dimensionless parameter $\hat{\sigma}$ is according to Carey [18] the accommodation coefficient which describes the rate for the liquid/vapor phase change. This parameter represents a multiplier on the phase change rate predicted by kinetic theory. Obviously Equations 2.53 and 2.54 depend strongly on the value of the accommodation coefficient $\hat{\sigma}$. According to Carey [18], for most engineering systems a value for $\hat{\sigma}$ of less then 1 is expected. No approved data are published for the accommodation coefficient of cryogenic fluids.

2.5 Thermal Stratification

The effect of thermal stratification in liquid cryogenic propellants results from external heat exchange, primarily through the side walls of the propellant tanks, and is defined as the development of temperature gradients within the liquid propellant. It is caused by the free-convection flow of heated liquid along the sidewalls of the tank and into the upper regions near the free surface. There, it flows toward the center of the tank, mixes and disperses resulting in a downward motion of the heated liquid. This forms a growing layer of thermally stratified liquid which is at higher temperature than the bulk. This layer is called thermal stratification layer or thermal boundary layer. The thickness of the thermal boundary layer δ_T can be determined analytically according to Baehr and Stephan [9] as

$$\delta_T \approx \sqrt{\pi D_{t,l}\, t} \tag{2.55}$$

with the liquid thermal diffusivity $D_{t,l}$ as defined in Equation 2.40 and the time t.

2.6 Pressure Drop Model

In the active-pressurization experiments presented in this study, a characteristic pressure drop appears after pressurization end. In the following section, a model is introduced for the analytical description of the pressure drop. The corresponding results are presented in Section 6.8.

Hopfinger and Das [54], Das and Hopfinger [27] introduced an analytical description of the rate of pressure drop in a closed cylindrical container due to liquid sloshing. This approach was revised by Ludwig et al. [66] and adapted to the analysis of the pressure drop due to sloshing in cryogenic liquids. The theoretical approach presented by Ludwig et al. [66] is transferred for this study to the subject of the pressure drop in a cryogenic tank after pressurization with evaporated propellant as pressurant gas. The derivation of the applied equation is presented hereafter:

According to Equation 2.26 follows the equation for the conservation of energy for liquids as

$$\rho_l c_{p,l} \left(\frac{\partial T_l}{\partial t} + (\boldsymbol{v_l} \bullet \nabla) T_l \right) = \lambda_l \nabla^2 T_l \tag{2.56}$$

with the liquid density ρ_l, the specific heat capacity of the liquid $c_{p,l}$, the liquid temperature T_l, the liquid velocity field $\boldsymbol{v_l}$ and the thermal conductivity λ_l. Incompressibility conditions are used. The experimental results, presented in Chapter 6 show that for the analysis of the pressure drop after active-pressurization, the liquid velocity can be neglected and only a vertical temperature gradient has to be considered. With the thermal diffusivity $D_{t,l}$ in a liquid, as defined in Equation 2.40 follows for Equation 2.56:

$$\frac{\partial T_l}{\partial t} = D_{t,l} \frac{\partial^2 T_l}{\partial z^2} \tag{2.57}$$

Integration of Equation 2.57 from $z = 0$ to the liquid height H_l results in the following equation.

$$\frac{\partial}{\partial t} \int_0^{H_l} T_l \, dz = D_{t,l} \left. \frac{\partial T_l}{\partial z} \right|_{z=H_l} \tag{2.58}$$

The experimental results presented in Section 6.2 show that no temperature gradient appears at $z = 0$. The maximal temperature difference Θ in the liquid is between the saturation temperature at the free surface T_{sat} and the liquid bulk temperature T_l. The thickness of the liquid temperature gradient corresponds approximately to the thickness of the thermal boundary layer δ_T which can be determined analytically with Equation 2.55. Consequently, it follows for Equation 2.58:

$$D_{t,l} \left. \frac{\partial T_l}{\partial z} \right|_{z=H_l} = D_{t,l} \frac{\Theta}{\delta_T} \tag{2.59}$$

The heat flux at the interface per unit surface $\dot{q}_\Gamma$ is derived from Fourier's law (Equation 2.38) as

$$\dot{q}_\Gamma = -\lambda_l \left. \frac{\partial T_l}{\partial z} \right|_{z=H_l} = -\rho_{l,sat} c_{p,l} D_{t,l} \left. \frac{\partial T_l}{\partial z} \right|_{z=H_l} \tag{2.60}$$

and the associated condensation heat flux is

$$\dot{q}_{\Gamma,cond} = v_{cond} \rho_{v,sat} \Delta h_v \tag{2.61}$$

with the vapor density at saturation conditions $\rho_{v,sat}$ and the specific enthalpy of evaporation Δh_v. With $\dot{q}_\Gamma = \dot{q}_{\Gamma,cond}$ and Equation 2.59, the vapor condensation velocity normal to the mean interface v_{cond} follows as

$$v_{cond}(t) = -\frac{D_{t,l} \, \rho_{l,sat}(t) \, c_{p,l}(t) \, \Theta(t)}{\rho_{v,sat}(t) \, \Delta h_v(t) \, \delta_T(t)} = -\frac{D_{t,l}}{\delta_T(t)} \frac{\rho_{l,sat}(t)}{\rho_{v,sat}(t)} \, \mathbf{Ja}(t) \tag{2.62}$$

where $\mathbf{Ja}$ is the JAKOB number $\mathbf{Ja} = (c_{p,l} \, \Theta)/\Delta h_v$.

The vapor phase is is considered as an ideal gas (see Section 2.7). Therefore, the pressure change in the system follows as

$$\frac{dp(t)}{dt} = \frac{R_s \bar{T}_{v,i}}{V_u} \frac{dm_v(t)}{dt} + \frac{p_i}{\bar{T}_{v,i}} \frac{d\bar{T}_v(t)}{dt} \tag{2.63}$$

where $\bar{T}_v$ is the ullage volume averaged mean vapor temperature, $\bar{T}_{v,i}$ is the initial mean vapor temperature and V_u is the ullage volume. The change of vapor mass is

$$\frac{dm_v(t)}{dt} = v_{cond}(t) \, \rho_{v,sat}(t) \, A_{cond}(t) + v_{evap}(t) \, \rho_{v,sat}(t) \, A_{evap}(t) \tag{2.64}$$

with A_{cond} as the area of the free surface where condensation takes place and A_{evap} where evaporation takes place with the evaporating vapor velocity v_{evap}. The pressure change as a function of time can be calculated using Equations 2.62, 2.63 and 2.64.

$$\frac{dp(t)}{dt} = \frac{R_s \bar{T}_{v,i} \, \rho_{v,sat}(t)}{V_u} \left[v_{evap}(t) A_{evap}(t) - \frac{\sqrt{D_{t,l}} \, \frac{\rho_{l,sat}(t)}{\rho_{v,sat}(t)} \, \mathbf{Ja}(t) \, A_{cond}(t)}{\sqrt{\pi \, t}} \right]$$
$$+ \frac{p_i}{\bar{T}_{v,i}} \frac{d\bar{T}_v(t)}{dt} \tag{2.65}$$

Integration of Equation 2.65 from pressurization end $t_{p,f}$ to the time t results in the following

$$p(t) = p_f \frac{\bar{T}_v(t)}{\bar{T}_{v,p,f}} - \frac{R_s \, \bar{T}_{v,p,f} \, \rho_{v,sat}(t)}{V_u} \left[\frac{2 \, \frac{\rho_{l,sat}(t)}{\rho_{v,sat}(t)} \, \mathbf{Ja}(t) A_{cond}(t) \sqrt{D_{t,l}}}{\sqrt{\pi}} \left(\sqrt{t} - \sqrt{t_{p,f}} \right) \right]$$
$$+ \frac{R_s \, \bar{T}_{v,p,f} \, \rho_{v,sat}(t)}{V_u} \left[v_{evap}(t) A_{evap}(t) \, (t - t_{p,f}) \right] \tag{2.66}$$

where $\mathbf{Ja}$, ρ and $\bar{T}_v$ are time dependent parameters and the initial mean vapor temperature at pressurization end $t_{p,f}$ is $\bar{T}_{v,p,f}$.

2.7 Scaling Concept for Active-Pressurization in Normal Gravity

Characteristic numbers represent a suitable scaling method of results achieved in laboratory size experiments for predictions for a full size application. In this section, the characteristic numbers which are relevant for the analysis of active-pressurization of cryogenic propellants are introduced and a comparison of their ranges for the performed sub-scale experiments of this study and a full scale application is given.

2.7.1 Derivation of Relevant Characteristic Numbers

In the following, the characteristic numbers, relevant for this study are derived from the conservation equations. The subsequent general assumptions are therefore made:

- A common liquid propellant tank is analyzed in normal gravity g, which is considered as two species two-phase system: the propellant, which consists of a liquid and a gaseous phase and the non-condensable pressurant gas, which is only regarded as gaseous phase. It is assumed that the amount of dissolved non-condensable gas in the liquid phase is small and can therefore be neglected in this study (see Section 6.5.1.2).

- Both liquid and gaseous phase are treated as a fluid with constant reference density ρ_{ref}. The Boussinesq approximation is applied, which assumes that effects based on volume expansions caused by temperature changes and species concentrations in the vapor phase can be neglected. This results in $\rho = \rho_{ref}$, except in the buoyancy term of the momentum equation.

- Viscosity μ, heat conductivity λ and heat capacity c_p are considered as constant. For the conservation of energy, kinetic effects can be neglected as well as viscous heating and heating due to body forces.

- The considered fluids are Newtonian fluids.

- The liquid phase of the propellant is treated as an incompressible fluid with a pressure independent density ρ_l. The thermodynamic pressure however, may change and therefore also the density.

- The reference density ρ_{ref} corresponds to the liquid or vapor density at norm pressure $p_{norm} = 101.3$ kPa with β_T being the coefficient for thermal expansion. This approximation is valid only in the vicinity of the liquid saturation temperature, where $T_{ref} \approx T_{sat}$.

- The thermal equation of state for fluids can be applied to the vapor phase (Equation 2.1).

- The liquid and the vapor phases are separated by a sharp interface with no velocity slip.

- The diffusions coefficients of liquid phase $D_{l,12} = D_{l,21}$ and vapor phase $D_{v,12} = D_{v,21}$, which are dependent on pressure and temperature, are handled as constants subsequent to one initial calculation.

- The temperature at the free surface is equal to the saturation temperature and has no temperature jumps (Baehr and Stephan [9]).

- No interface resistance for the phase change as well as no velocity slip at walls are presumed.

- Dissipation is not taken into account within the energy conservation equation.

- Heat transfer by radiation inside the tank is neglected.

In the following, the conservation equations introduced in Section 2.2 with the above mentioned assumptions are applied for a propellant tank as depicted in Figure 2.1. The liquid and vapor phase are considered separately. The vapor phase is treated as a two species system. The liquid phase is assumed as one species system and the validation of this assumption will be presented in Section 6.5.1.2.

The liquid phase is considered as a pure substance. Consequently, it follows from Equations 2.12, 2.18 and 2.26 for the liquid phase

$$\nabla \bullet \boldsymbol{v}_l = 0 \tag{2.67}$$

$$\rho_{l,ref} \left(\frac{\partial \boldsymbol{v}_l}{\partial t} + (\boldsymbol{v}_l \bullet \nabla)\boldsymbol{v}_l \right) = -\nabla p_l + \mu_l \nabla^2 \boldsymbol{v}_l + \rho_l \, \boldsymbol{g} \tag{2.68}$$

$$\rho_{l,ref} \, c_{p,l} \left(\frac{\partial T_l}{\partial t} + (\boldsymbol{v}_l \bullet \nabla)T_l \right) = \lambda_l \nabla^2 T_l. \tag{2.69}$$

For the vapor phase follows accordingly for species 1

$$\frac{\partial \rho_{v,1}}{\partial t} + \boldsymbol{v}_v \bullet \nabla \rho_{v,1} = D_{v,12} \nabla^2 \rho_{v,1} \tag{2.70}$$

and for species 2

$$\frac{\partial \rho_{v,2}}{\partial t} + \boldsymbol{v}_v \bullet \nabla \rho_{v,2} = D_{v,21} \nabla^2 \rho_{v,2}. \tag{2.71}$$

With the species-specific mass fraction Ψ_i from Equation 2.4 and the composition coefficient for the vapor $\beta_{\Psi,v}$ defined as

$$\beta_{\Psi,v} = -\frac{1}{\rho_v} \frac{\partial \rho_v}{\partial \Psi_{v,1}} \tag{2.72}$$

follows for the vapor density ρ_v

$$\rho_v = \rho_{v,ref}(1 - \beta_{T,v}(T_v - T_{v,ref}) - \beta_{\Psi,v}(\Psi_{v,1} - \Psi_{v,ref})). \tag{2.73}$$

The Boussinesq approximation is applied to consider the temperature and composition dependency of the gaseous density ρ_v in correlation to the gaseous reference density $\rho_{v,ref}$ and temperature $T_{v,ref}$ with the thermal expansion coefficient $\beta_{T,v}$ and the composition coefficient $\beta_{\Psi,v}$. It therefore follows for the conservation of momentum for the vapor phase:

$$\rho_{v,ref} \left(\frac{\partial \boldsymbol{v}_v}{\partial t} + (\boldsymbol{v}_v \bullet \nabla)\boldsymbol{v}_v \right) = -\nabla p_v + \mu_v \nabla^2 \boldsymbol{v}_v + \rho_v \, \boldsymbol{g} \tag{2.74}$$

The equation for the conservation of energy for the vapor phase follows as:

$$\rho_{v,ref} \, c_{p,v} \left(\frac{\partial T_v}{\partial t} + (\boldsymbol{v}_v \bullet \nabla)T_v \right) = \lambda_v \nabla^2 T_v \tag{2.75}$$

The following boundary conditions are considered: Between the liquid phase and the adjacent wall, the heat flux density based on the heat conductivity of the liquid phase is defined using Fourier's law (Equation 2.38)

$$\dot{q}_{l,w} = -\lambda_l \frac{\partial T_l}{\partial \boldsymbol{N}_w} \bigg|_w \bullet \boldsymbol{n}_w = \alpha_{l,w}(T_w - \bar{T}_l) \tag{2.76}$$

where $\boldsymbol{n}$ is the dimensionless normal unit vector of the surface and $\boldsymbol{N}$ is the normal unit vector of the surface with one unit of length. For the vapor phase and the corresponding tank wall and lid follows accordingly

$$\dot{q}_{v,w} = -\lambda_v \frac{\partial T_v}{\partial \boldsymbol{N}_w}\bigg|_w \cdot \boldsymbol{n}_w = -\alpha_{v,w}(T_w - \bar{T}_v). \tag{2.77}$$

At the liquid surface, the normal velocity of the phase boundary can be written as:

$$\boldsymbol{v}_\Gamma \cdot \boldsymbol{n}_\Gamma = \boldsymbol{v}_{l,\Gamma} \cdot \boldsymbol{n}_\Gamma - \frac{\hat{m}_\Gamma}{\rho_l} = \boldsymbol{v}_{v,\Gamma} \cdot \boldsymbol{n}_\Gamma - \frac{\hat{m}_\Gamma}{\rho_v} \tag{2.78}$$

with $\hat{m}_\Gamma$ as the mass flux at the free surface.

The boundary condition for the velocity jump at the phase boundary is

$$\Delta \boldsymbol{v}_\Gamma = (\boldsymbol{v}_{l,\Gamma} \cdot \boldsymbol{n}_\Gamma - \boldsymbol{v}_{v,\Gamma} \cdot \boldsymbol{n}_\Gamma)\boldsymbol{n}_\Gamma = \frac{\rho_v - \rho_l}{\rho_l \rho_v}\hat{m}_\Gamma \boldsymbol{n}_\Gamma \tag{2.79}$$

and the propellant mass fraction in the vapor phase follows as:

$$\hat{m}_{\Gamma,1} = \Psi_{v,1}\hat{m}_\Gamma - D_{v,12}\frac{\partial \rho_{v,1}}{\partial \boldsymbol{N}_\Gamma} \cdot \boldsymbol{n}_\Gamma \tag{2.80}$$

The jump in the conductive heat flux $\Delta \dot{q}_\Gamma$ at the phase interface, dependent on the mass flux $\hat{m}_\Gamma$, is given by the following correlation.

$$\Delta \dot{q}_\Gamma = \hat{m}_\Gamma \Delta h_{v,1} = -\left(\lambda_l \frac{\partial T_{l,\Gamma}}{\partial \boldsymbol{N}_\Gamma} - \lambda_v \frac{\partial T_{v,\Gamma}}{\partial \boldsymbol{N}_\Gamma}\right) \cdot \boldsymbol{n}_\Gamma \tag{2.81}$$

For the dimensionless formulation, the conservation equations from Section 2.2 are scaled by the following characteristic quantities: The reference density is ρ_{ref}, the reference temperature is T_{ref}, the characteristic velocity is v_{ref}, the characteristic length scale is L, the characteristic time is τ, the characteristic acceleration is g, the characteristic temperature difference is Θ and the characteristic mass flux is J. Consequently, the characteristic hydrodynamic pressure is $\rho_{ref} v_{ref}^2$ according to the stagnation pressure.

In the following, the dimensionless parameters are indicated by asterisk symbols and are defined as

$$\begin{aligned} \boldsymbol{v}^* = \frac{\boldsymbol{v}}{v_{ref}}; \quad p^* = \frac{p}{\rho_{ref}\, v_{ref}^2}; \quad T^* = \frac{T - T_{ref}}{\Theta}; \quad t^* = \frac{t}{\tau}; \\ \boldsymbol{g}^* = \frac{\boldsymbol{g}}{g}; \quad j^* = \frac{j}{J}; \hat{m}^* = \frac{\hat{m}}{J}; \quad \nabla^* = \nabla\, L. \end{aligned} \tag{2.82}$$

It therefore follows from the Conservation Equations 2.67, 2.68 and 2.69 for the liquid phase, that

$$\nabla^* \cdot \boldsymbol{v}_l^* = 0 \tag{2.83}$$

$$\left(\frac{L}{\tau \, v_{ref}}\right)\frac{\partial \boldsymbol{v}_l^*}{\partial t^*} + (\boldsymbol{v}_l^* \cdot \nabla^*)\boldsymbol{v}_l^* =$$

$$-\nabla^* p_l^* + \left(\frac{\mu_l}{\rho_{l,ref} \, v_{ref} \, L}\right)\nabla^{*2}\boldsymbol{v}_l^* + \left(\frac{g \, L}{v_{ref}^2}\right)\boldsymbol{g}^* - \left(\frac{g \, \beta_{T,l} \, \Theta \, L}{v_{ref}^2}\right)T_l^*\boldsymbol{g}^* \tag{2.84}$$

$$\left(\frac{L}{\tau \, v_{ref}}\right)\frac{\partial T_l^*}{\partial t^*} + \boldsymbol{v}_l^* \cdot \nabla^* T_l^* = \left(\frac{\lambda_l}{\rho_{l,ref} \, c_{p,l} \, v_{ref} \, L}\right)\nabla^{*2}T_l^*. \tag{2.85}$$

and for the ullage phase accordingly results from Equations 2.70, 2.71, 2.74 and 2.75.

$$\left(\frac{L}{\tau \, v_{ref}}\right)\frac{\partial \Psi_{v,1}}{\partial t^*} + \boldsymbol{v}_v^* \cdot \nabla^* \Psi_{v,1} = \left(\frac{D_{v,12}}{v_{ref} \, L}\right)\nabla^{*2}\Psi_{v,1} \tag{2.86}$$

$$\left(\frac{L}{\tau \, v_{ref}}\right)\frac{\partial \Psi_{v,2}}{\partial t^*} + \boldsymbol{v}_v^* \cdot \nabla^* \Psi_{v,2} = \left(\frac{D_{v,21}}{v_{ref} \, L}\right)\nabla^{*2}\Psi_{v,2} \tag{2.87}$$

$$\left(\frac{L}{\tau \, v_{ref}}\right)\frac{\partial \boldsymbol{v}_v^*}{\partial t^*} + (\boldsymbol{v}_v^* \cdot \nabla^*)\boldsymbol{v}_v^* = -\nabla^* p_v^* + \left(\frac{\mu_v}{\rho_{v,ref} \, v_{ref} \, L}\right)\nabla^{*2}\boldsymbol{v}_v^* + \left(\frac{g \, L}{v_{ref}^2}\right)\boldsymbol{g}^*$$

$$-\left(\frac{g \, \beta_{T,v} \, \Theta \, L}{v_{ref}^2}\right)T_v^* \, \boldsymbol{g}^* - \left(\frac{\beta_{\Psi_v} \, g \, L}{v_{ref}^2}\right)(\Psi_{v,1} - \Psi_{v,ref}) \, \boldsymbol{g}^* \tag{2.88}$$

$$\left(\frac{L}{\tau \, v_{ref}}\right)\frac{\partial T_v^*}{\partial t^*} + \boldsymbol{v}_v^* \cdot \nabla^* T_v^* = \left(\frac{\lambda_v}{\rho_{v,ref} \, c_{p,v} \, v_{ref} \, L}\right)\nabla^{*2}T_v^* \tag{2.89}$$

The boundary conditions of the heat flux density in the liquid phase at the tank wall can therefore be written as

$$\left.\frac{\partial T_l^*}{\partial \boldsymbol{n_w}}\right|_w \cdot \boldsymbol{n}_w = -\frac{\alpha_{w,l}L}{\lambda_l}(T_w^* - \bar{T}_l^*) \tag{2.90}$$

and for the vapor phase

$$\left.\frac{\partial T_v^*}{\partial \boldsymbol{n_w}}\right|_w \cdot \boldsymbol{n}_w = -\frac{\alpha_{w,v}L}{\lambda_v}(T_w^* - \bar{T}_v^*). \tag{2.91}$$

The normal velocity of the phase boundary can be written as

$$\boldsymbol{v}_\Gamma^* \cdot \boldsymbol{n}_\Gamma = \boldsymbol{v}_{l,\Gamma}^* \cdot \boldsymbol{n}_\Gamma - \frac{J}{\rho_l v_{ref}}\hat{m}_\Gamma^* = \boldsymbol{v}_{v,\Gamma}^* \cdot \boldsymbol{n}_\Gamma - \frac{J}{\rho_v v_{ref}}\hat{m}_\Gamma^* \tag{2.92}$$

and the velocity jump at the phase boundary follows as:

$$\Delta \boldsymbol{v}_\Gamma^* = (\boldsymbol{v}_{l,\Gamma}^* \cdot \boldsymbol{n}_\Gamma - \boldsymbol{v}_{v,\Gamma}^* \cdot \boldsymbol{n}_\Gamma)\boldsymbol{n}_\Gamma = -\left(\frac{\rho_l}{\rho_v} - 1\right)\frac{J}{\rho_l v_{ref}}\hat{m}_\Gamma^*\boldsymbol{n}_\Gamma \tag{2.93}$$

The propellant mass fraction in the vapor phase can be written as

$$\hat{m}_{\Gamma,1}^* = \Psi_{v,1}\hat{m}_\Gamma^* - \frac{D_{v,12}\rho_{v,ref}}{LJ}\frac{\partial \Psi_{v,1}}{\partial \boldsymbol{n}_\Gamma} \cdot \boldsymbol{n}_\Gamma \tag{2.94}$$

and the jump in the conductive heat flux at the phase boundary follows as:

$$\hat{m}_\Gamma^* = -\left(\frac{\lambda_l \Theta}{\Delta h_{v,1}JL}\frac{\partial T_{l,\Gamma}^*}{\partial \boldsymbol{n}_\Gamma} - \frac{\lambda_v \Theta}{\Delta h_{v,1}JL}\frac{\partial T_{v,\Gamma}^*}{\partial \boldsymbol{n}_\Gamma}\right) \cdot \boldsymbol{n}_\Gamma \tag{2.95}$$

Implementing the relevant characteristic numbers, summarized in Table 2.1, leads to the nondimensional form of the conservation equations of the liquid:

$$\nabla^* \bullet v_l^* = 0 \tag{2.96}$$

$$\mathbf{St}\,\frac{\partial v_l^*}{\partial t^*} + (v_l^* \bullet \nabla^*)v_l^* = -\nabla^* p_l^* + \frac{1}{\mathbf{Re}}\nabla^{*2}v_l^* + \frac{1}{\mathbf{Fr}}\,g^* - \frac{1}{\mathbf{Fr}_{\beta_T}}\,T_l^*g^* \tag{2.97}$$

$$\mathbf{St}\,\frac{\partial T_l^*}{\partial t^*} + v_l^* \bullet \nabla^*T_l^* = \frac{1}{\mathbf{Re\,Pr}}\nabla^{*2}T_l^* \tag{2.98}$$

The nondimensional form of the conservation equations of the ullage follows accordingly:

$$\mathbf{St}\frac{\partial \Psi_{v,1}}{\partial t^*} + v_v^* \bullet \nabla^*\Psi_{v,1} = \frac{1}{\mathbf{Re_v\,Sc_v}}\nabla^{*2}\Psi_{v,1} \tag{2.99}$$

$$\mathbf{St}\frac{\partial \Psi_{v,2}}{\partial t^*} + v_v^* \bullet \nabla^*\Psi_{v,2} = \frac{1}{\mathbf{Re_v\,Sc_v}}\nabla^{*2}\Psi_{v,2} \tag{2.100}$$

$$\begin{aligned}
\mathbf{St}\,\frac{\partial v_v^*}{\partial t^*} + (v_v^* \bullet \nabla^*)v_v^* = &-\nabla^* p_v^* + \frac{1}{\mathbf{Re_v}}\nabla^{*2}v_v^* + \frac{1}{\mathbf{Fr}}\,g^* - \frac{1}{\mathbf{Fr}_{\beta_T v}}\,T_v^*g^* \\
&- \frac{1}{\mathbf{Fr}_\Psi}\,(\Psi_{v,1} - \Psi_{v,ref})g^*
\end{aligned} \tag{2.101}$$

$$\mathbf{St}\,\frac{\partial T_v^*}{\partial t^*} + v_v^* \bullet \nabla^*T_v^* = \frac{1}{\mathbf{Re_v\,Pr_v}}\nabla^{*2}T_v^* \tag{2.102}$$

For the boundary condition results the following: the heat flux density in the liquid phase at the tank wall is

$$\left.\frac{\partial T_l^*}{\partial n_w}\right|_w \bullet n_w = -\mathbf{Nu}(T_w^* - \bar{T}_l^*) \tag{2.103}$$

and for the vapor phase at the tank wall follows:

$$\left.\frac{\partial T_v^*}{\partial n_w}\right|_w \bullet n_w = -\mathbf{Nu_v}(T_w^* - \bar{T}_v^*) \tag{2.104}$$

The normal velocity of the phase boundary is

$$v_\Gamma^* \bullet n_\Gamma = v_{l,\Gamma}^* \bullet n_\Gamma - \frac{\mathbf{Ev\,Ja}}{\mathbf{Re\,Pr}}\hat{m}_\Gamma^* = v_{v,\Gamma}^* \bullet n_\Gamma - \frac{\mathbf{Ev_v\,Ja_v}}{\mathbf{Re_v\,Pr_v}}\hat{m}_\Gamma^* \tag{2.105}$$

and the velocity jump at the phase boundary follows as:

$$\Delta v_\Gamma^* = (v_{l,\Gamma}^* \bullet n_\Gamma - v_{v,\Gamma,v}^* \bullet n_\Gamma)n_\Gamma = -(\mathbf{R} - 1)\frac{\mathbf{Ev\,Ja}}{\mathbf{Re\,Pr}}\hat{m}_\Gamma^* n_\Gamma \tag{2.106}$$

The propellant mass fraction in the vapor phase is

$$\hat{m}_{\Gamma,1}^* = \Psi_{v,1}\hat{m}_\Gamma^* - \frac{\mathbf{Pr_v}}{\mathbf{Sc_v\,Ev_v\,Ja_v}}\frac{\partial \Psi_{v,1}}{\partial n_\Gamma} \bullet n_\Gamma \tag{2.107}$$

and the jump in the conductive heat flux at the phase boundary is

$$\hat{m}_\Gamma^* = -\left(\frac{1}{\mathbf{Ev}}\frac{\partial T_{l,\Gamma}^*}{\partial n_\Gamma} - \frac{1}{\mathbf{Ev_v}}\frac{\partial T_{v,\Gamma}^*}{\partial n_\Gamma}\right) \bullet n_\Gamma. \tag{2.108}$$

Table 2.1 Relevant characteristic numbers

$\mathbf{Ev} = \dfrac{\Delta h_v \; J \; L}{\lambda_l \; \Theta}$	Evaporative heat flux number
$\mathbf{Ev_v} = \dfrac{\Delta h_v \; J \; L}{\lambda_v \; \Theta}$	Evaporative heat flux number vapor phase
$\mathbf{Fr} = \dfrac{v_{ref}^2}{g \; L}$	FROUDE number
$\mathbf{Fr}_{\beta_T} = \dfrac{v_{ref}^2}{g \; \beta_{T,l} \; \Theta \; L}$	Thermal expansion FROUDE number
$\mathbf{Fr}_{\beta_T \mathbf{v}} = \dfrac{v_{ref}^2}{g \; \beta_{T,v} \; \Theta \; L}$	Thermal expansion FROUDE number vapor phase
$\mathbf{Fr}_{\Psi} = \dfrac{v_{ref}^2}{g \; \beta_{\Psi} \; \Theta \; L}$	Species expansion FROUDE number
$\mathbf{Ja} = \dfrac{c_{p,l} \; \Theta}{\Delta h_v}$	JAKOB number
$\mathbf{Ja_v} = \dfrac{c_{p,v} \; \Theta}{\Delta h_v}$	JAKOB number vapor phase
$\mathbf{Nu} = \dfrac{\alpha L}{\lambda_l}$	NUSSELT number
$\mathbf{Nu_v} = \dfrac{\alpha L}{\lambda_v}$	NUSSELT number vapor phase
$\mathbf{Pr} = \dfrac{\mu_l \; c_{p,l}}{\lambda_l}$	PRANDTL number
$\mathbf{Pr_v} = \dfrac{\mu_v \; c_{p,v}}{\lambda_v}$	PRANDTL number vapor phase
$\mathbf{R} = \dfrac{\rho_l}{\rho_v}$	Density ratio
$\mathbf{Re} = \dfrac{\rho_{l,ref} \; v_{ref} \; L}{\mu_l}$	REYNOLDS number
$\mathbf{Re_v} = \dfrac{\rho_{v,ref} \; v_{v,ref} \; L}{\mu_v}$	REYNOLDS number vapor phase
$\mathbf{Sc_v} = \dfrac{\mu_v}{\rho_{v,ref} \; D_{i,j}}$	SCHMIDT number vapor phase
$\mathbf{St} = \dfrac{L}{\tau \; v_{ref}}$	STROUHAL number

2.7.2 Scaling Ranges

The most important dimensionless numbers characterizing active-pressurization of cryogenic propellant tanks are considered hereafter. The presented experimental setup is scaled down from the upper stage concept intended for Europe's future launcher concept Adapted Ariane 5 ME. For the relevant characteristic numbers, scaling between the experiments, described in this study, and the full-scale upper stage concept is presented. For the experiments, liquid nitrogen (LN2) is used as model propellant, pressurized with gaseous nitrogen (GN2) or gaseous helium (GHe) of different temperatures. The experimental tank is described in detail in Section 4.1. The considered full-scale cryogenic upper stage has a liquid hydrogen (LH2) tank with a volume of approximately 64.5 m^3, pressurized with gaseous hydrogen (GH2), as well as a liquid oxygen (LOX) tank with a volume of approximately 21.8 m^3, which is pressurized with gaseous helium. It is a Common Bulkhead tank configuration with the LH2 tank having an upper dome radius of 3 m and being situated above the LOX tank, which has an upper dome radius of 2 m. The convex Common Bulkhead is in the direction of fight. The tank pressure after pressurization ranges for the LN2 experiments of this study between 200 kPa and 400 kPa and the tank pressure of the full-scale LH2 tank lies between 200 kPa and 350 kPa and for the full-scale LOX tank between 200 kPa and 300 kPa.

The JAKOB number is a very important dimensionless number for this active-pressurization study. It is derived from the boundary conditions at the phase interface and represents the correlation between the sensible heat and the latent heat of evaporation Δh_v, describing the heat transfer at phase change. According to Table 2.1, the JAKOB number of the liquid phase is defined as

$$\mathbf{Ja} = \frac{c_{p,l}\, \Theta_l}{\Delta h_v} \tag{2.109}$$

with $c_{p,l}$ being the specific heat capacity of the liquid at constant volume and Θ_l the characteristic temperature difference for the liquid phase, which is defined for this study as

$$\Theta_l = T_{sat} - T_{ref}. \tag{2.110}$$

T_{ref} is the reference temperature, which is the saturation temperature at the norm pressure $p_{norm} = 101.3$ kPa. For nitrogen, the reference temperature is 77.35 K, for oxygen it is 90.19 K and for parahydrogen 20.27 K. T_{sat} is the saturation temperature at the current tank pressure. This parameter is selected for the characteristic temperature difference as the tank pressure is a key parameter for the active tank pressurization, which is restricted by

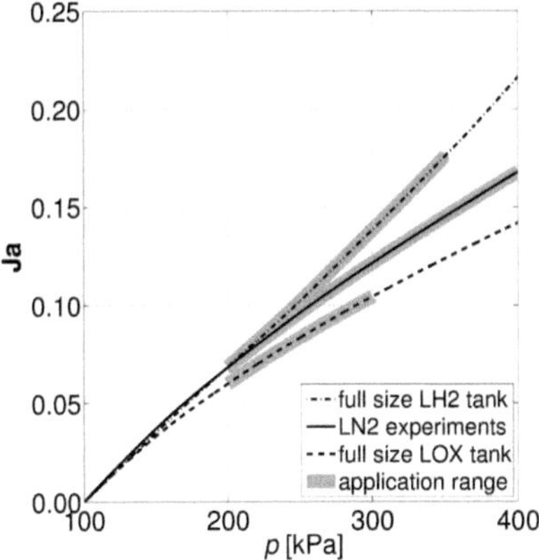

Fig. 2.2 Evolution of the JAKOB number over the tank pressure for the liquid phase of the LN2 experiments and full size LH2 and LOX tanks.

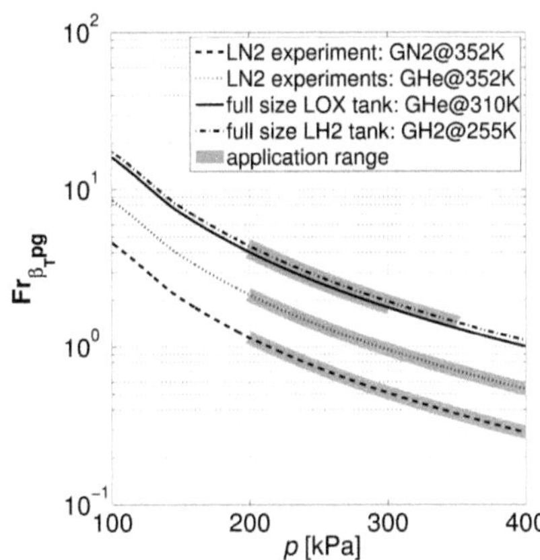

Fig. 2.3 Evolution of the thermal expansion FROUDE number of the pressurant gas over the tank pressure for two LN2 experiments and full size LH2 and LOX tanks. The corresponding pressurant gas temperatures are given.

propellant, structural and engine requirements. The fluid parameters are calculated for the saturation points at the current tank pressures. Figure 2.2 shows the JAKOB number over the tank pressure for the liquid phase of the experiments as well as for the full size LOX and LH2 upper stage propellant tanks. The gray highlighted areas are the applied pressure ranges: for the LN2 experiments, the final tank pressures are between 200 kPa and 400 kPa. The tank pressure of the full-scale LH2 tank lies between 200 kPa and 350 kPa and for the LOX tank between 200 kPa and 300 kPa. The JAKOB number is only dependent on pressure and temperature and is not time-dependent. Figure 2.2 shows that the values for the JAKOB number lie between $0.06 \leq \mathbf{Ja} \leq 0.18$, which indicates thermodynamic similarity of the fluids.

Another important dimensionless number for the active-pressurization is the thermal expansion FROUDE number. For the analysis of thermal convection, it represents forced convection at the ratio to natural convection. For this study, the thermal expansion FROUDE number of the pressurant gas is of interest. It can therefore be written in the following way.

$$\mathbf{Fr}_{\beta_T\mathbf{pg}} = \frac{v_{pg,ref}^2}{g\,\beta_{T,pg}\,\Theta_{pg}\,L} \tag{2.111}$$

Where g is the gravitational acceleration and $\beta_{T,pg}$ is the thermal expansion coefficient of the pressurant gas. The characteristic temperature difference for the pressurant gas is chosen for this study as the difference between the maximal pressurant gas inlet temperature $T_{pg,m}$ and the reference temperature T_{ref}.

$$\Theta_{pg} = T_{pg,m} - T_{ref} \tag{2.112}$$

The maximal pressurant gas inlet temperature $T_{pg,m}$ is applied, as the pressurant gas temperature is an important parameter for the active-pressurization, especially for the cryogenic propellants. The characteristic length L is defined by the aspect ratio with the ullage volume V_u over the area of the free surface A_Γ.

$$L = \frac{V_u}{A_\Gamma} \tag{2.113}$$

The reference velocity of the pressurant gas $v_{pg,ref}$ is defined as the pressurant gas mass flow rate $\dot{m}_{pg}$ over the diffuser's pressurant gas outlet area A_{dif} and the density of the pressurant gas ρ_{pg}.

$$v_{pg,ref} = \frac{\dot{m}_{pg}}{A_{dif}\ \rho_{pg}} \tag{2.114}$$

For the experiments, the pressurant gas outlet area is the cylinder jacket area of the diffuser (see Section 4.1). The dimensions of the diffuser can be found in Figure A.1 in the Appendix. For the experiments, the bottom of the diffuser is sealed to ensure only radial outflow. This results in a pressurant gas outlet area A_{dif} of $3.24 \cdot 10^{-4}$ m^2 for the LN2 experiments. To enable the comparison of the experiments and the full-scale upper stage, also for the LOX and the LH2 tanks a cylindrical diffuser is assumed with its pressurant gas outlet area at the cylinder jacket. With a diameter of 0.3 m and a height of 0.15 m follows a the pressurant gas outlet area of 0.14 m^2. For the LN2 experiments with GHe as pressurant gas, the pressurant gas inlet mass flow rate is $1.62 \cdot 10^{-4}$ kg/s and for the experiments, pressurized with GN2, the inlet mass flow rate is $8.32 \cdot 10^{-4}$ kg/s. The characteristic length for the experiments is calculated with the ullage volume of $V_u = 0.014$ m^3 and the free surface area of $A_\Gamma = 0.0688$ m^2 as $L = 0.2$ m. Concerning the full-scale LOX tank with GHe as pressurant gas, a pressurant gas inlet mass flow rate of 0.095 kg/s is assumed, a maximal pressurant gas temperature of 310 K, and a characteristic length of 0.17 m results from an assumed ullage volume of 0.65 m^3 (3 % of the tank volume) and a resulting initial free surface area of 3.8 m^2. For the full-scale LH2 tank with GH2 as pressurant gas, a pressurant gas inlet mass flow rate of 0.077 kg/s, a pressurant gas temperature of 255 K and a characteristic length of $L = 0.21$ m are used. The characteristic length results from an assumed ullage volume of 1.5 m^3 (2.3 % of the tank volume) and a resulting initial free surface area of 7.1 m^2.

The thermal expansion FROUDE number is used in this study to evaluate the influence of the pressurant gas inlet temperature on the active-pressurization process. For the presented experiments, two different pressurant gases are evaluated: gaseous nitrogen, as evaporated propellant pressurization and gaseous helium as inert gas pressurization. Figure 2.3 shows the thermal expansion FROUDE number of the pressurant gas over the tank pressure for two LN2 experiments, one with GN2 as pressurant gas with an inlet temperature of 352 K and one with GH2 as pressurant gas, also with an inlet temperature of 352 K. These data are compared to the thermal expansion FROUDE number of the pressurant gas of the gas full-scale LOX and LH2 tanks. For the LOX tank, GHe at 310 K is considered as inlet temperature and for the LH2 tank GH2 at 255 K. The gray highlighted areas are the applied pressure ranges. In Figure 2.3, it can be seen that the thermal expansion FROUDE numbers of the pressurant gas for the LOX and LH2 tanks are quite close, whereas the experimental data are two and four times smaller. One main reason is the very high pressurant gas inlet velocity applied for the full-scale tanks of up to 4.4 m/s for GHe and 5.8 m/s for GH2 at ambient pressure. Due to technical reasons, the maximal feasible inlet velocity of the experiments was 3.7 m/s for GHe and 2.7 m/s for GN2 at ambient pressure. The inlet velocity decreases with increasing tank pressure due to the increasing fluid density.

The REYNOLDS number represents the ratio of inertial forces to viscous forces. Turbulent flow occurs at high REYNOLDS numbers and is therefore dominated by inertial forces. Laminar flow occurs at low REYNOLDS numbers where the viscous forces are dominant. It is defined for the pressurant gas as

$$\mathbf{Re_{pg}} = \frac{\rho_{pg} v_{pg,ref} L}{\mu_{pg}} = \frac{v_{pg,ref} L}{\nu_{pg}} \tag{2.115}$$

with μ_{pg} being the dynamic viscosity, ρ_{pg} the density and ν_{pg} the kinematic viscosity of the pressurant gas. For the full scale LOX tank with GHe as the pressurant gas, a pressurant gas inlet mass flow rate of 0.095 kg/s is determined with a diffuser area of 0.14 m^2, a characteristic length of 0.17 m and a maximal pressurant gas temperature of 310 K. This results in a REYNOLDS number of $\mathbf{Re_{pg}} = 5.7 \cdot 10^3$. With the data already presented for the thermal expansion FROUDE number, a REYNOLDS number for the full-scale LH2 tank with GH2 as the pressurant gas of $\mathbf{Re_{pg}} = 1.5 \cdot 10^4$ results. For the LN2 experiments pressurized with GN2,

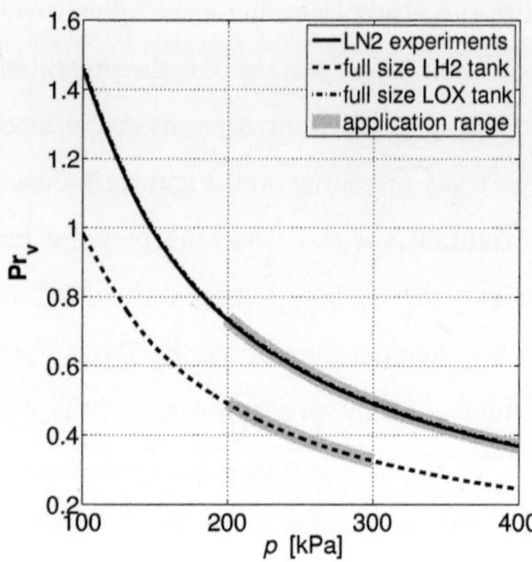

Fig. 2.4 Evolution of the PRANDTL number for the tank ullage over the tank pressure for the LN2 experiments and full size LH2 and LOX tanks.

the maximal REYNOLDS number is determined as $\mathbf{Re_{pg,m}} = 5.4 \cdot 10^4$. For the LN2 experiments with GHe as pressurant gas appears a maximal REYNOLDS number of $\mathbf{Re_{pg,m}} = 5.6 \cdot 10^3$. This indicates similarity between the full-scale and the sub-scale pressurant gas flow.

Another important characteristic number for this study is the PRANDTL number. The PRANDTL number represents the ratio of kinematic viscosity to thermal diffusivity. It describes the correlation between the momentum and the heat transfer and is defined for the vapor phase inside the propellant tanks as

$$\mathbf{Pr_v} = \frac{c_{p,v}\,\mu}{\lambda} = \frac{\nu_v\,\rho_{v,ref}\,c_{p,v}}{\lambda_v}$$
$$= \frac{\nu_v}{D_{t,v}} \tag{2.116}$$

with λ_v as the thermal conductivity of the vapor. The thermal diffusivity $D_{t,v}$ is applied in the form $D_{t,v} = \lambda_v/(\rho_{v,ref}c_{p,v})$ with the reference density $\rho_{v,ref} = \rho_v$ at norm pressure $p_{norm} = 101.3$ kPa.

The PRANDTL number is only dependent on the fluid and the fluid state. Figure 2.4 depicts the PRANDTL number for the tank ullage of the experiments as well as for the LOX and LH2 full scale tanks. The ullage is assumed to be exclusively filled with the respective evaporated propellant and the average ullage temperatures are 142 K for the LN2 experiments, 110 K for the LOX, and 40 K for the LH2 full-scale tanks. The gray highlighted areas show the application range of the tank pressure. The PRANDTL number differs in this range from $0.33 \leq \mathbf{Pr_v} \leq 0.74$ and the ullage PRANDTL number of the LN2 experiments corresponds to that of the full size LH2 tank.

For the pressurant gas, the PRANDTL number is calculated with the respective fluid parameters.

$$\mathbf{Pr_{pg}} = \frac{\nu_{pg}\,\rho_{ref,pg}\,c_{p,pg}}{\lambda_{pg}} \tag{2.117}$$

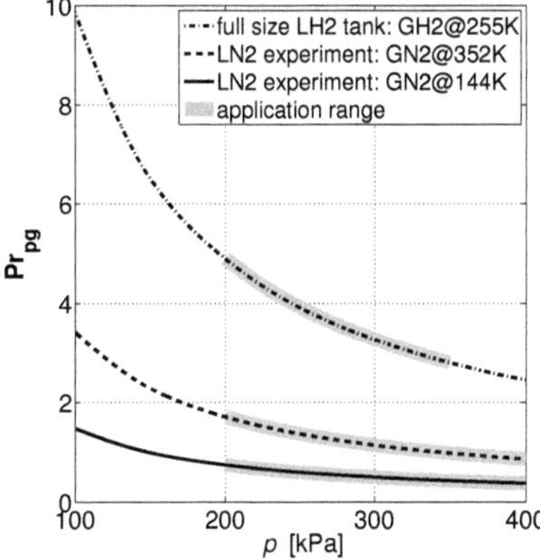

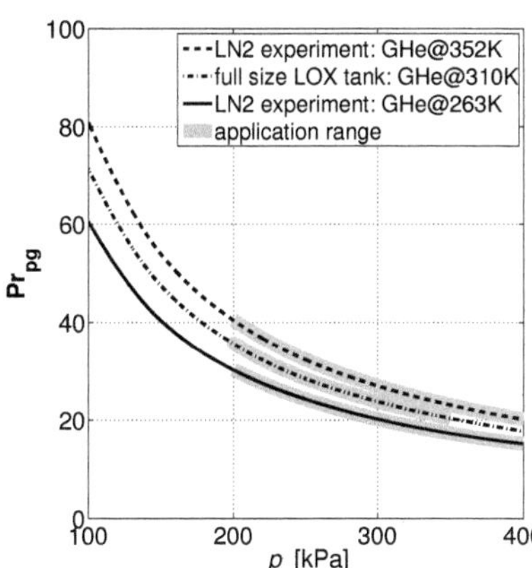

Fig. 2.5 Evolution of the PRANDTL number for the evaporated-propellant pressurant gas over the tank pressure for two LN2 experiments pressurized with GN2 and for the full size LH2 tank with GH2 as pressurant gas. The corresponding pressurant gas temperatures are given.

Fig. 2.6 Evolution of the PRANDTL number for the pressurant gas helium over the tank pressure for two LN2 experiments and a full size LOX tank. The corresponding pressurant gas temperatures are given.

The relating graphs for the PRANDTL number of the pressurant gas are depicted in Figure 2.5. The upper stage LH2 tank, pressurized with GH2 at 255 K, is compared to the LN2 experiments with the pressurant gas GN2 at 352 K and 144 K, as they are used in the experiments. Figure 2.6 shows the PRANDTL number for GHe as pressurant gas. The presented pressurant gas temperatures for the LN2 experiments are 352 K and 263 K. For the full-scale LOX tank 310 K is assumed as pressurant gas temperature. For the evaporated-propellant pressurization, the PRANDTL numbers lie between $0.37 < \mathbf{Pr_{pg}} < 4.9$ and for helium as pressurant gas between $15.2 < \mathbf{Pr_{pg}} < 40.5$.

The NUSSELT number $\mathbf{Nu}$ describes the ratio of convective to conductive heat transfer across a boundary surface. This characteristic number is used in this study for the determination of the mode of heat transfer between the tank wall and the fluids in Section 6.6.2. It is however assumed to be not essential for the scaling of the results in this study. The STROUHAL number $\mathbf{St}$ describes oscillating flow mechanisms and is therefore not very relevant for this study. The SCHMIDT number $\mathbf{Sc}$ is defined as the ratio of the viscous diffusion rate to mass diffusivity, characterizing fluid flows with simultaneous momentum and mass diffusion convection processes. As for the active-pressurization process, forced convection has a greater influence on the mixing of the pressurant gas and the propellant vapor than the rate of mass diffusivity, also this dimensionless number is not significant for this study.

Chapter 3
STATE OF RESEARCH

This chapter will present an overview of the current state of research in the field of cryogenic propellant tank pressurization. The relevant literature is considered for analytical approaches, numerical considerations, experimental work and cognate disciplines. The objectives of this study are subsequently given. Literature referring to pressurization in low gravity conditions are explicitly excluded from the survey, as it is not subject of the study.

The majority of scientific research relating to cryogenic propellant tank pressurization was carried out in the 1960s in the USA. Driven by the Apollo program, numerous analytical, numerical and experimental research works were performed. During that time, Ring [80] published a monograph about rocket propellant and pressurization systems. It gives an overview on pressurization systems, ullage requirements for pressure control systems, pressurant gas thermodynamics and pressurization hardware requirements. Another monograph, published by NASA (National Aeronautics and Space Administration) [72] in 1975 is exclusively dedicated to pressurization systems for liquid rockets. It includes a review and discussion of the design of the pressurization systems and represents a summary of existing design criteria and techniques, as well as recommended practices.

3.1 Analytical Approaches

A large number of studies presented in the 1960s focus on the development of analytical methods for the preliminary design of pressurization systems. Clark [22] classified the applied analytical approaches for the description of the cryogenic propellant pressurization as *distributed systems* and *lumped systems*. For a distributed system, temperature, composition, velocity or pressure in the tank ullage is determined as a function of space and time. The calculations are based on the conservation equations of mass, energy and momentum. This analytical approach has the advantage that it is the most general and usually provides the greatest amount of information concerning the pressurization process. However, the approach of the distributed system requires often parameters and terms, of which is generally little known a priori. One-dimensional analytical approaches of this kind can be found for the analysis of the pressurized discharge of a liquid from a container, for example in Arpaci et al. [7], Arpaci and Clark [6], Coxe and Tatom [25], Momenthy [69] and Clark et al. [23].

In a lumped or nodal system, only the mean properties of e.g. the temperature of the gas space or the tank wall are determined. With this approach, the vapor temperature is not a function of position, as the mean gas temperature is calculated as function of time only. The lumped system can be used for the determination of overall property changes. These analyses are thermodynamic in nature and provide a minimum amount of information. Gluck and Kline [46] introduced in 1962 a study on pressurant gas requirements, using an empirical method for the calculation of the pressurization rate of a cryogenic propellant tank. Further lumped-system studies for propellant tank pressurization during discharge can be found in the work of Moore et al. [70], Bowerstock and Reid [16], Humphrey [55], and Bizjak et al. [14].

At the department of Space Launcher Systems Analysis (SART) at the DLR Institute of Space Systems, the in-house program PMP (Propellant Management Program) is used for the preliminary design of liquid propellant management systems for launcher and launcher related objects. PMP is a system analysis tool, based on the lumped-system approach and is used for preliminary design studies within a launch system context. The program is validated with data from existing launcher stages. Currently, PMP is reviewed and extended for the assessment of the propellant management during ballistic phases and the more accurate sizing of the propellant tank pressurization systems for cryogenic launcher stages. Based on the nominal required propellant mass, the mission timeline, the engine characteristics and a few more inputs, PMP sizes the propellant tanks with the total required propellant mass, the feed systems as well as the pressurization system. PMP has been applied in numerous system studies. Recently performed studies were presented by Dumont et al. [36, 37] and Ludwig and Dreyer [65].

In 2012, Kim et al. [57] published an analytical study on the active-pressurization of the liquid oxygen (LOX) tank with gaseous helium (GHe) of the Korean Space Launch Vehicle KSLV-I. In this study, the pre-launch pressurization phase and the subsequent readiness check phase (without pressurization) are analyzed. They assessed the influence of the final tank pressure and the ullage volume on the temperature and pressure drop after pressurization end. The results are summarized below:

- The pressure drop after pressurization end is nearly independent of the final tank pressure.
- An increasing pressurization level results in a fairly linear increasing amount of evaporating LOX.
- The required helium mass increases linearly with the pressurization level.
- A larger volume of the initial tank ullage leads to a smaller pressure drop after pressurization end.
- The volume of the initial tank ullage has negligible influence on the evaporation process during the pressurization phase. After pressurization end, however the amount of evaporated LOX significantly increases with increasing tank ullage.
- Absorption of GHe into LOX during pressurization is negligible.

3.2 Numerical Considerations

As early as the 1960s, attempts to develop and apply numerical methods to the analysis and the design of cryogenic propellant pressurization systems were intensively pursued by major American aerospace companies and NASA centers. In 1965, Clark [22] and Epstein et al. [40] presented their work on the "Rocketdyne Program", a finite-difference code for the propellant tank pressurization analysis developed at the American aerospace company Rocketdyne. This program was updated by Nein and Thompson in 1966, as experimental data for large scale systems became available (Nein and Thompson [75], Thompson and Nein [96]).

Based on computations with this program, Epstein [38] developed a generalized equation for the prediction of the total required pressurant gas mass for LOX and liquid hydrogen (LH2) propellant tanks pressurized with either GHe or its propellant vapor. The total required pressurant gas mass is often referred to as "pressurant requirements" in NASA literature. However, this correlation by Epstein [38] is limited to the analysis of pressurization of cylindrical tanks in combination with liquid expulsion. Furthermore, it is based on the concept of the collapse factor, which is defined as the ratio of total pressurant requirements to the pressurant requirements under conditions of zero heat and mass transfer. Further studies, dedicated to the concept of the collapse factor were published by de Quay [78], de Quay and Hodge [79] and van Dresar [30]. These studies showed that the collapse factors varied widely depending on the cryogenic propellant, the tank size and configuration, and pressurization system. Models to analytically determine the collapse factor showed varied levels of accuracy compared to empirical collapse factors.

In 1968, Epstein and Anderson [39] extended the Epstein equation [38], based on the theory of Gluck and Kline [46] and Arpaci et al. [7] and data points from the Rocketdyne program. This extended equation is also applying the concept of the collapse factor and applicable for any axisymmetric propellant tank containing any cryogenic propellant. In 1997, van Dresar [31] published a revised version of the correlation of Epstein and Anderson [39] and compared it to test data from spherical propellant tanks by Stochl et al. [90, 91, 92, 93] and van Dresar and Stochl [34]. However, this correlation still has the disadvantage of being limited to the prediction of the pressurant requirement in combination with propellant expulsion and is not applicable for pressurization only.

In parallel to the Rocketdyne program, Roudebush [81] developed a similar code at the Lewis Research Center of the NASA which was applied in 1966 by DeWitt et al. [29] to compare results from that model with experimental data. The code has subsequently been revised by Stochl et al. [90, 91, 92, 93] and Masters [67], who presented in 1974 a separate version of the code for the pressurization phase, also often called ramp phase. Besides that numerical consideration, Roudebush and Mandell [82] presented in 1965 an analytical investigation of the pressurized LH2 tank outflow problem. They introduced the approach of using modified wall and gas Stanton numbers for the determination of the required pressurant gas mass during outflow and compared their approach to experimental data.

The numerical Rocketdyne and the Roudebush codes have in common that they solve the continuity and energy equations only in the axial direction for a cylindrical coordinate system. It was found that this one-dimensional modeling is often appropriate for axisymmetric propellant tanks in normal or high acceleration levels. However, due to the assumptions of the programs, the pressurant gas is required to be properly diffused, which excludes the application for straight pipe injector systems. Epstein et al. [40] stated that by using a diffuser for the pressurant gas, the dominating energy exchange resulted between the tank ullage and the adjacent tank wall. Epstein et al. also claimed that the heat and mass transfer over the free surface should have an considerable effect on the total required pressurant gas mass only for high pressurant gas inlet temperatures and small tank sizes. Roudebush [81] noted that free convection correlations work well with gaseous hydrogen and helium as pressurant gases. When using heavier molecular weight gases, such as oxygen, better results were obtained with the Rocketdyne code by specifying combined free and forced convection for the gas-to-wall heat transfer (Nein and Thompson [75]).

The Rocketdyne and the Roudebush programs are generally constrained to one-dimensional considerations and are based on simplifying assumptions and experimentally derived correlations. The proper modeling of the pressurization process requires simultaneous solving the fluid-mechanical, thermodynamical and heat transfer models for a tank with two phases of matter. With the advancements in computing power and the enhancement of numerical simulation methods of the last decades, modeling of the propellant tank pressurization process has been extended with multidimensional computational techniques, such as CFD (computational fluid dynamics). In 1990, Hardy and Tomsik [51] presented their work on the prediction of the ullage gas thermal stratification in a NASP (National Aero-Space Plane) LH2 tank using Flow-3D, a commercial CFD software. This code is able to model compressible flow with a two-fluid interface, thermal buoyancy, surface tension, turbulent flow and time dependent boundary conditions (for more details see Section 5.1). Part of the work of Hardy and Tomsik [51] was the determination of the vapor temperature profiles and pressurant gas requirements during the ramp pressurization process. They found that it was the quantity of gas added and not the ramp rate that affected the magnitude of the ramp profiles. However, they only modeled the tank ullage and the liquid surface was set as a rigid plane with constant temperature. On that account, no mass and the associated heat transfer over the free surface was considered.

In 1991, Sasmal et al. [85] presented a study on the computational modeling of the pressurization process in a spherical hydrogen tank, also using Flow-3D. They compared their numerical results to the experimental results of Stochl et al. [92]. Similar to Hardy and Tomsik [51], Sasmal et al. [85] modeled the liquid hydrogen surface as a solid boundary with a specified heat flux and therefore disregarded phase change at the free surface. Moreover, a single heat flux rate at the tank wall was specified. It was found, that both assumptions were probably incorrect. In addition, they found that the pressurization time and gas requirement is nearly independent of the initial ullage temperature distribution. In 1993, Sasmal et al. [84] presented further work with Flow-3D on the influence of heat transfer rates and pressurant gas mass flow rates on pressurization of a LH2 propellant tank with GH2 as pressurant gas. In this study, the specified heat flux, used in [85] was replaced by the division of the ullage tank wall into several azimuthal strips, each defined as separate obstacle, for which the temperature was computed. They found that the ullage boundary heat flux rates do significantly effect the pressurization process. It was stated that minimizing the heat loss from the ullage and maximizing the pressurant gas flow rates results in the minimization of the required pressurant gas mass. Also for this study, no mass transfer at the free surface was considered.

Grayson et al. also used Flow-3D for the analysis of cryogenic propellant behavior, however they did not analyze the active-pressurization phase. Only the following topics were addressed: thermal stratification in connection with sloshing (Grayson and Navickas [50], Grayson [47]), self-pressurization (Grayson et al. [49], Grayson et al. [48]), as well as an axial jet thermodynamic vent system (Lopez et al. [64]).

In 2000, Adnani and Jennings [1] presented a study on pressurization analysis of cryogenic propulsion systems using the commercial CFD code FLUENT. Experimental data from a LH2 tank pre-pressurized with helium and during draining with gaseous hydrogen (GH2) was used for the validation of the modeling methodology. They stated, that due to existing limitations of the FLUENT code, the liquid phase inside the tank could not be modeled and it was therefore assumed to have a constant temperature, equal to the initial liquid temperature. For the determination of the heat transfer to the liquid phase, pre-existing wall function routines were used. The applied code did not have the capability to account for phase change. It was therefore neglected and assumed to have no significant impact on the tank pressurization. The pressurization curve of the simulation was stated to be in excellent agreement with that of the experiment. It was pointed out, that during the GHe pressurization, a part of the helium shows recirculation patterns in the tank near the diffuser and some helium mixes with the GH2. They stated that heat transfer between the warm GHe and the colder GH2 through forced convection appears, as well as heat transfer between the helium and the tank wall, on which it impinges and then flows along. During the GH2 pressurization after GHe pre-pressurization, the added GH2 tends to accumulate toward the top of the tank. Moreover, the tank ullage was found to be stratified, although the experiment neglected external heat fluxes due to the use of a vacuum chamber facility. They also found that the diffuser design has an impact on the pressurization curve.

Wang et al. [99, 100, 101] published in 2013 three studies on the numerical investigation of pressurization performance in combination with propellant draining using FLUENT. The developed numerical model was validated with the experimental data from Roudebush [81]. In [99], they focused on the heat exchange inside the tank and outside due to aerodynamic heating. A LOX tank pressurized with GHe was analyzed and for the calculations, phase change was assumed to be negligible and the vapor phase only consisted of GHe. It was found that the aerodynamic heating effect for the considered tank with a foam insulation system can be disregarded for a pressurization analysis. Moreover, the wall to foam heat exchange was found to have a certain influence on the pressurization prediction. In [100], focus was placed on the effects of various influencing factors on the pressurization with draining. A LH2 tank, pressurized either with GH2 or GHe was therefore analyzed. They stated that an increased pressurant gas temperature and a thin tank wall can reduce the pressurant gas requirements. Furthermore, a straight pipe injector can also reduce pressurant gas mass but results in a remarkable evaporation of the propellant. In [101], the same LH2 tank was investigated, either pressurized with GH2 or GHe. They stated that the influence of phase change on the pressurization performance during a GH2 pressurized discharge test case can be disregarded, when using a foam-insulated tank. Moreover, it was found that during pressurized discharge, GH2 has a better pressurization performance than GHe due to its higher specific heat capacity.

In 2008, Ahuja et al. [2] presented a comprehensive numerical framework, using multi-element unstructured CFD and rigorous real fluid property routines, for the analysis of propellant tank and delivery systems. This framework represented an extension of a in-house compressible gas/liquid code CRUNCH CFD of Combustion Research and Flow Technology, Inc. In the presented study, different simulations of cryogenic propellant tank effects were performed. An active-pressurization case was also included, where a spherical LOX tank was pressurized with gaseous oxygen (GOX). They stated, that for the presented test case, significant mixing in the vapor phase and phase change appears during active-pressurization. Mattick et al. [68] presented a hybrid "sharp interface" CFD procedure, which combines the attributes of the CRUNCH CFD procedure with a multi-node lumped parameter approach. An internal boundary separated the tank ullage and liquid phase. This approach was applied to an active-pressurization experiment of an LH2 tank with ambient temperature helium under normal gravity, described by Barnett et al. [10]. The results for the pressure rise were stated to be in excellent agreement with the experiment. Additionally it was found that strong thermal stratification appears in the vapor phase.

In 2012, Kwon et al. [58] published their work on modeling the prediction of helium mass requirement for propellant tank pressurization during draining. They presented a numerical model for the helium mass prediction of the pressurization of a propellant tank using SINDA/FLUINT, a general purpose thermal/fluid network analyzer. They made the following assumptions: an axial temperature stratification is present in the tank ullage and the referring tank wall, the liquid surface temperature remains at the saturation temperature corresponding to the tank pressure before pressurization, the tank ullage is initially filled with the same gas as the pressurant gas. The latter was stated as the reason for the assumption that mass transfer can be disregarded, as the non-condensable gas helium was used as pressurant gas for all analyzed test cases. The model was verified by data from cryogenic propellant drainage ground tests with liquid nitrogen (LN2) as model propellant, pressurized with GHe, as well as in flight-test data of a sounding rocket with kerosene and LOX as propellants, pressurized with GHe. Furthermore, Kwon et al. performed a parametric study as sensitivity test of the developed model. It was found that the heat transfer coefficient between the ullage and the tank wall has the greatest effect on the accuracy of the model. Modelling the heat transfer as natural convection was stated as appropriate. In a further parametric study, the effect of the supplied helium temperature on the required helium mass was analyzed. An inversely proportional correlation was found although it was stated that no heat transfer occurs between the tank ullage and its surroundings. When the heat transfer was considered, the heat transfer to the tank walls was found to increase with increasing helium temperature. This required again more helium to achieve the required tank pressure.

Leuva et al. [63] presented in 2012 a study on the initial pressurization of a LH2 tank with GHe with focus on the pressure collapse during engine ignition. A numerical model was introduced using ANSYS CFX, a commercial CFD program, where only the tank ullage and the tank wall was considered and the LH2 free surface was substituted with a wall with constant temperature. The pressurization was performed by pumping a pulsating flow of hot GHe into the tank. It was found that the pressure collapse appears when no more pressurant gas is injected and a decreased temperature of the tank walls increases the pressure drop.

3.3 Experimental Work

In addition to the analytical and numerical considerations a large number of experiments were performed for the analysis of the cryogenic propellant tank pressurization. In most cases, the experimental data was also used for the validation of the developed models.

In 1960, van Wylen et al. [97] and Fenster et al. [42] presented experimental data on pressurization without propellant discharge of boiling LN2 with GN2 as pressurant gas. Van Wylen et al. [97] presented data for the evolution of the temperature of one thermocouple in the liquid phase and one at the wall in the liquid phase. They also logged the boil-off rate. It was found that after pressurization end the liquid temperature and wall temperature increases linearly until a quasi steady-state point, where the liquid temperature reaches the new saturation temperature. They stated that the boil-off is the result of heat transfer from the wall into the liquid near the liquid surface. Moreover, 85% of the heat transferred across the wall is used to raise the liquid temperature until the time when the liquid temperature reaches the new saturation temperature. It should be noted that for both studies the tank pressure evolution during the experiment remains unclear.

In the same year, Bowersock et al. [15] presented experimental data of horizontal LN2 and LOX tanks, pressurized with gaseous nitrogen (GN2) or GOX. They found, that the estimation of gas requirements is not sensitive to changes in pressurization rate or initial temperature in the tank wall. They stated that for the use of condensable pressurant gas, a diffuser is necessary to minimize turbulence of the liquid and the subsequent condensation losses.

In 1962, Gluck and Kline [46] also presented experimental results for the validation of their analytical relation. Data from discharge experiments from a LH2 tank were introduced, using either ambient-temperature GH2 or GHe as pressurant gas. The tank was pressurized and in the next step, propellant outflow was initiated and pressurant gas was supplied continually to keep the tank pressure constant. They stated that the pressurant weight can be reduced by introducing the pressurant gas at temperatures substantially greater than that of LH2. Important results of the discharge experiments were that for GH2-pressurized discharge the appearing type of mass transfer is evaporation rather then condensation, whereas for helium-pressurized discharge, evaporation is the dominating type of mass transfer. It was stated however that the effect of mass transfer on the pressurant requirements was found to be negligibly small. Moreover, the major heat transfer encounters from the vapor phase to the adjacent tank wall.

Clark [24] performed pressurization experiments with LN2 and GN2 as the pressurant gas, using a floating styrofoam piston on the free surface. He stated that during pressurization, a condensate film appeared on the piston. The heat transfer from the free surface of the film through the piston was found to be similar to the transient heat conduction in a semi-infinite slab.

In 1962, Nein and Head [74] presented experimental data from the pressurized discharge of a full-scale Saturn liquid oxygen system (pressurized with GOX), a large single LOX tank (pressurized with GOX) and a small model LN2 tank (pressurized with either GN2 or GHe). They found that for the full-scale tanks, considerable mass transfer appears during expulsion between the gas in the tank ullage and the liquid propellant. The appearing mass transfer was always evaporation, except for the test, where the small model LN2 tank was pressurized with GN2. In that test, condensation occurred during the entire discharge test. Furthermore, the major internal heat transfer was stated to take place between the ullage gas and the liquid surface, and that the heat transfer between the ullage gas and the adjacent tank walls are of minor importance. However, the results for the small model tank indicated a reversal in the predominant mode of heat transfer.

Nein and Thompson [75] presented in 1966 experimental results for the pressurization of five large scale tank configurations. This data was used for the advancement of the Rocketdyne program. Pressurization data from cylindrical and spheroidal tanks, ranging in size over four orders of magnitude were presented. As results were summarized, that also in large scale propellant tanks, no significant radial ullage temperature gradient occurs. The heat transfer between the vapor and tank walls can differ significantly from free convection, depending on the tank geometry and the diffuser design. However, the strongest influence on pressurant weight has the pressurant gas inlet temperature. Further important influencing factors are the tank radius, distributor flow area and aerodynamic heating. An analysis of the mass transfer showed that condensation occurs for all of the evaluated test cases, except for one test with a high pressurant gas inlet temperature. The axial ullage temperature gradients were linear when oxygen was used as a pressurant gas and concave for helium pressurization, indicating evaporation as mass transfer. Another interesting result was the analysis of the helium concentration in the tank ullage: for experiments in a LOX tank, pre-pressurized with helium and pressurized with gaseous oxygen during expulsion, the maximum helium concentration appeared approximately in the middle of the ullage height, gradually decreasing in both directions. For tests, exclusively pressurized with helium, GHe filled the main part of tank ullage. Only above the free surface, GOX existed with decreasing concentration from the free surface to the tank top.

At around the same time, Olsen [76] presented an analytical and experimental investigation of interfacial heat and mass transfer in a pressurized LH2 tank. He conducted experiments with ambient-temperature GH2 or GHe as pressurant gases. After ramping, the tank pressure was held constant by continuous injection of the pressurant gas. Olsen stated that the condensation, occurring at the GH2 pressurized tests, creates an appreciable heat transfer to the liquid phase which must be considered in thermal stratification studies. Furthermore, he found that the effect of interfacial heat transfer on the pressurant requirements could probably be neglected; however, the effect of interfacial mass transfer should be considered.

Another topic, which was experimentally investigated is the influence of the injector geometry on the pressurant gas requirements. DeWitt et al. [29] presented in 1966 an experimental evaluation of the effect of the pressurant gas injector geometry on the required pressurant gas mass for a cylindrical LH2 tank, pressurized with GH2. Six injector geometries (cone, hemisphere, disk, radial, multiple screen, straight pipe) were analyzed with different pressurant gas inlet temperatures. They found that all analyzed diffuser type injectors have similar pressurant requirements during the overall tank cycle, which consisted of ramp, hold and expulsion time periods. The greatest added pressurant gas mass was for the multiple screen geometry and was the least for the cone injector. For the straight pipe injectors however, a significant decrease in pressurant gas requirements during the expulsion period was found, as less heat is transferred to the tank walls and a greater amount of LH2 evaporates. Furthermore, for the straight pipe injectors it was found that the amount of evaporation increases with decreasing ramp time due to the more forceful impingement of the pressurant gas on the liquid surface. For the diffuser-type injectors, during the ramp period mass transfer occurs generally in the form of condensation.

Stochl et al. [90, 91, 92, 93] presented in 1969 and 1970 four reports on experimental investigations on the tank pressurization and expulsion of LH2 from two different spherical tanks. Gaseous hydrogen and gaseous helium were used as pressurant gases. The objective was to determine the effect of various physical parameters on the pressurant gas requirements during the initial pressurization and expulsion phases. The analyzed parameters were the outflow rate, pressurizing rate, initial ullage volume, pressurant gas temperature, injector geometry, tank wall thickness and final tank pressure level. The experimental results were compared to results from the revised and extended Roudebush code. The work of Stochl et al. [90, 91, 92, 93] has the following important outcomes for the initial tank pressurization:

- Increasing the inlet gas temperature results in a decrease in the pressurant requirement for constant ramp rates.
- Increasing the ramp rate results in a decrease in the pressurant requirement for constant inlet gas temperatures.

- Larger initial ullage volumes at constant ramp rates result in increased pressurant requirement.
- Increased ramp rates at given initial ullage volumes lead to decreased pressurant gas requirements.

In Stochl et al. [90, 91, 93], they stated that the results of the modified Roudenbush code were adequate to allow the prediction of the approximate pressurant gas requirements during the initial pressurization using both pressurant gases and a diffuser type injector. For the initial GH2 pressurization of the smaller tank however, the code failed to accurately predict the pressurant requirements, particularly for small initial ullage volumes [92]. It was determined for all experiments that the inlet gas temperature has the strongest influence on the pressurant gas requirements, closely followed by the injector design.

In 1970, Lacovic [60] presented results of ramp and expulsion tests of a LOX tank, similar to that of the Centaur stage, pressurized with helium. The experimental results for the helium requirements were compared to the results of calculations performed with the Roudebush code. The performed pressurization tests consisted of the ramp period and the hold time, at which the tank pressure was kept constant and a small amount of oxygen was discharged for the chilldown of the engine pumps and the propellant feed system. It was observed that the helium requirements increases with increasing tank ullage, however the increase is not directly proportional to the tank ullage due to the outside heat input. No significant variation in helium requirements with ramp time appears.

Based on the work of Stochl et al. [90, 91, 92, 93], DeWitt and McIntire [28] presented in 1974 results of pressurant expulsion test for the discharge of liquid methane, in order to determine the effect of various physical parameters on the pressurant gases methane, helium, hydrogen and nitrogen. The experimental results were compared to the results of the revised Roudebush code. It was seen that the use of gaseous methane as a pressurant gas results in significant condensation and that gaseous nitrogen has high solubility in methane. On that account, the analytical program could not be used for the GN2 expulsions. Furthermore, it was stated that for GHe and GH2 as pressurant gases, the inlet gas temperature had a negligible effect on the gas requirements.

Van Dresar and Stochl [34] presented in 1993 experimental data of the pressurization and expulsion of a full-scale LH2 tank, pressurized with GH2. They stated that the mass transfer rate plays a significant role in ramp phases of practical duration. It was found that during a three minute ramp phase with a pressurant gas temperature of 280 K, mass transfer was not constant, but switches from evaporation to condensation and back to evaporation. Additionally it was stated that the pressurant requirements increase with increasing ramp time and that the largest portion (for long ramp periods 50% or more) of the input energy goes into wall heating.

3.4 Cognate Disciplines

This section introduces fields of research that are related to the subject of active cryogenic propellant tank pressurization, but are not the focus of this study.

Sloshing

Firstly, the subject of sloshing of liquid propellants needs to be considered. Sloshing of propellant in launcher tanks is in general an undesired effect, as excessive sloshing can ultimately lead to mission failure. As examples for general research in this field, the work of Royon-Lebeaud et al. [83], Hopfinger and Das [54] and Das and Hopfinger [27] should be mentioned. Sloshing of cryogenic fluids was analyzed for example by Lacapere et al. [59], Moran et al. [71], Arndt [3] and van Foreest [45], who performed numerical studies based on the experimental data presented by Arndt [3]. One important effect of sloshing in cryogenic propellant tanks is a drop in pressure. Numerous experiments have been performed to understand this effect and in the majority of the test cases, an active-pressurization process was applied as a preparatory phase for the sloshing experiments. In the studies of Lacapere et al. [59], Moran et al. [71], Arndt [3] and van Foreest [45], active-pressurization was also used to reach the desired operating tank pressure. In the studies of Lacapere et al. [59], Arndt [3], and van Foreest [45], shaking of the test tank was immediately initiated after the final tank pressure was reached due to active-pressurization. In Moran et al. [71], a brief hold phase with constant tank pressure followed the ramp phase, after which sloshing was initiated. As a result of sloshing, a pressure drop appeared in all mentioned experiments. As it will be presented in this study, a pressure drop also occurs after pressurization end without sloshing. In order to further understand the situation in the propellant tank prior to sloshing, Ludwig et al. [66] performed cryogenic sloshing experiments with the following procedure: the tank pressure was ramped up from ambient to the final tank pressure using active-pressurization. Afterwards, the pressurization was stopped and the expected pressure drop due to the pressurization end appeared. As it will be presented in Section 6.8, the pressure drop after active-pressurization end shows an asymptotic evolution. Sloshing was therefore initiated when the tank pressure showed a nearly horizontal evolution. This procedure ensured that the subsequent pressure drop is only due to sloshing. The results for the sloshing phase of these experiments are not the subject of this study and can be found in Ludwig et al. [66]. However, the pressurization and relaxation phase is considered closely in this study in order to understand the initial conditions before sloshing.

Self-Pressurization

Another subject that should be mentioned here is the self-pressurization in a closed cryogenic tank. Ambient heat input heats the cryogenic liquid up, resulting in evaporation at the liquid surface and therefore into an increase of the tank pressure. Self-pressurization of cryogenic fluids is a field of research on its own and not subject of this study. Numerous analytical, numerical and experimental studies have previously been performed. The most spectacular self-pressurization experiment was the orbital flight of the Saturn IB vehicle AS-203 in 1966. This flight also included a closed fuel tank experiment, where the LH2 tank was closed to determine the pressure rise due to self-pressurization in orbit [26]. As a result of that experiment, the common bulkhead to the LOX tank ruptured and the vehicle exploded. Ahuja et al. [2] used an in-house CFD program and Grayson et al. [49] applied Flow-3D for the computational analysis of this self-pressurization experiment. Other examples for self-pressurization experiments are the work of Seo and Jeong [88], who presented experimental results of the self-pressurization of a LN2 tank for various heat leaks; Hasan et al. [53], who analyzed experimentally the self-pressurization of a full-scale LH2 tank subjected to low heat flux; Aydelott [8] who presented various self-pressurization test on a spherical LH2 dewar; Arnett and Millhiser [4], who introduced an analytical method for analyzing self-pressurization in a cryogenic fluid container due to side wall heating; Arnett and Voth [5], who presented a computer program for the calculation of thermal stratification and self-pressurization in a closed liquid hydrogen tank; Barsi and Kassemi [12], who compared numerical self-pressurization results of the commercial CFD code FLUENT with experimental of a LH2 tank and van Dresar et al. [33], who analyzed the effects of the fill level at low heat flux on the self-pressurization in a full-scale LH2 tank.

Thermal Stratification

The effect of thermal stratification in liquid cryogenic propellants results from external heat input, primarily through the side walls of the propellant tanks, and is defined as the development of temperature gradients within the liquid propellant. It is caused by the free-convection flow of heated liquid along the sidewalls of the tank and into the upper regions near the free surface. There, the warm liquid flows toward the center of the tank, mixes and disperses resulting in a downward motion of the heated liquid. This forms a growing layer of stratified liquid which is at higher temperature than the bulk. This layer is called thermal stratification layer or thermal boundary layer δ_T. Thermal stratification in liquid cryogenic propellants is a whole field of research on its own. However, thermal stratification should be considered in the analysis of active-pressurization of cryogenic propellant tanks. Some examples of literature relating to this topic are named here: The monograph of Ring [80] includes a whole chapter on the cryogenic propellant stratification analysis and Barnett et al. [10] investigated stratification in a full-scale LH2 tank of the Saturn S-IV stage. Clark [22] presented a review of stratification phenomena in cryogenic liquids and Segel [86] presented experimental results for the evolution of the thermal stratification after pressurization end in a LH2 container pressurized with helium under different values of heat influx. Arnett and Millhiser [4] presented an analytical method for the determination of the extent and severity of thermal stratification of a side wall heated cryogenic propellant container and the associated self-pressurization. Chin et al. [20] presented an analytical and experimental study of stratification in standard and reduced gravity fields and introduced in [21] an analytical and experimental study on the stratification within the context of liquid-ullage coupling, the thermodynamic interaction between the liquid and the vapor phase. In 1972, Arnett and Voth [5] presented a computer program for the calculation of thermal stratification and self-pressurization in a closed liquid hydrogen tank and in 1988, Navickas [73] presented a numerical study using Flow-3D for the analysis of thermal stratification in a liquid tank.

Ullage Mixing

In 1971, Kendle [56] analyzed analytically the influence of ullage mixing effects of active pressurization with a diffuser-type injector on the pressurization performance. He applied a 1-D model and found that ullage mixing changes the gas temperature profile and therefore the gas-wall temperature difference and heat transfer rate. He advised an extensive mixing of the pressurant gas and the tank ullage for reduction of the gas-wall heat loss and therefore a reduction of the pressurant requirements as it would be possible with a straight-pipe injector.

3.5 Objective of this Study

For efficient design and operation of cryogenic propellant tank pressurization systems, accurate predictions of pressurant gas requirements are needed: too little amount of pressurant gas could ultimately lead to the failure of the mission. On the other hand too much on-board pressurant gas mass results in considerable weight and cost penalties and, which is of the utmost importance for launchers, less payload mass. The majority of the active-pressurization studies carried out in the past were in combination with propellant discharge. This study focuses only on the initial active-pressurization phase prior to engine ignition and does therefore not consider propellant expulsion. The detailed analysis of this specific phase is very important for the understanding of the complete pressurization period, as it represents the preparatory phase, which determines the initial conditions of the subsequent phases. In the review of the current state of research it became clear that there are still several uncertainties in the field of active-pressurization of cryogenic propellant tanks. Challenges for the exact estimation of the pressurant requirements for the initial active-pressurization of cryogenic propellants are various: for example the amount of heat and mass transfer over the free surface, the mode of mass transfer, the effect of the pressurant gas injection temperature and diffuser design and location, the influence of the ullage volume, the evolution of thermal stratification in the fluids, the pressure drop after pressurization end, the absorption of helium in the liquid phase, the influence of the ramp rate, the pressurant gas flow pattern and the type of convection (free or forced) in the tank ullage and therefore the amount of the gas-to-wall heat transfer.

Accurate predictions concerning the active-pressurization phase for cryogenic propellants for the launcher applications require an improved understanding of the complex fluid-dynamic and thermodynamic phenomena. With today's experimental possibilities, numerical codes and computing power, it is feasible to further develop the knowledge on this area. The objective of this study is therefore to improve the understanding of the thermodynamic and fluid-dynamic phenomena of cryogenic propellant tank pressurization for the launcher application. The initial active-pressurization phase prior to engine ignition is specifically considered. Of interest is active-pressurization with both condensable and non-condensable pressurant gases. On this account, the following subjects are selected to be analyzed in more detail in this study:

- Pressure and temperature evolution
- Thermal stratification
- Required pressurant gas mass
- Phase change
- Heat transfer
- Pressure drop after pressurization end

Ground experiments were performed in a sub-scale propellant tank with liquid nitrogen as cryogenic model propellant, pressurized with gaseous nitrogen or gaseous helium. The preliminary LOX/LH2 upper stage design of the Adapted Ariane 5 ME launcher concept was applied as a scaling model. Numerical simulations were performed using the commercial CFD software Flow-3D. Analytical approaches and correlations were also applied. In this study, the results of the experiments, numerical simulations and analytical considerations and correlations are described and discussed and conclusions drawn from these results are presented.

Chapter 4
EXPERIMENTAL SETUP FOR ACTIVE-PRESSURIZATION

The motivation of this study is to further understand the thermodynamic and fluid dynamic effects, present during the active-pressurization process in order to optimize the on-board pressurant gas mass for future launchers. Experiments were therefore performed that focused on the investigation of the initial active-pressurization process. Special emphasis was placed on the influence of the pressurant gas inlet temperature on the required pressurant gas mass. For the experiments, liquid nitrogen (LN2) was used as a cryogenic propellant substitute, which was actively pressurized under normal gravity conditions up to different final tank pressures. Gaseous nitrogen (GN2) or gaseous helium (GHe) were used as the pressurant gas, with different inlet temperatures. The following chapter describes the experimental setup and the test procedure.

4.1 Experimental Setup

This section contains the description of the test hardware for the active-pressurization experiments of this work. Specifically, the active-pressurization test facility, the instrumentation, the required fluid properties and the heat input are described. The applied test facility was previously used by Sionkiewicz [89] for the analysis of heat transfer in cryogenic liquids. For the study presented here, some modifications of the setup were necessary to bring the active-pressurization process into focus. Arndt [3]

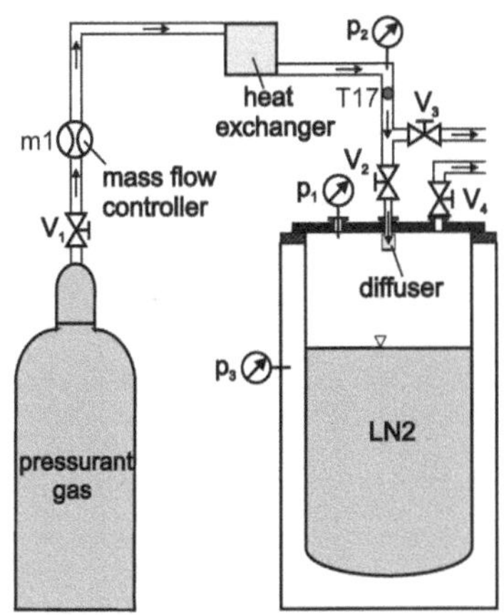

Fig. 4.1 Test setup with the pressurant gas storage bottle, mass flow controller, heat exchanger, the high pressure tank, corresponding lines, valves and pressure sensors.

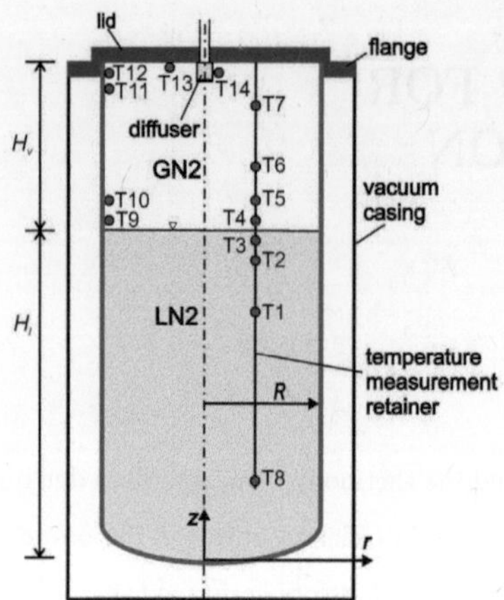

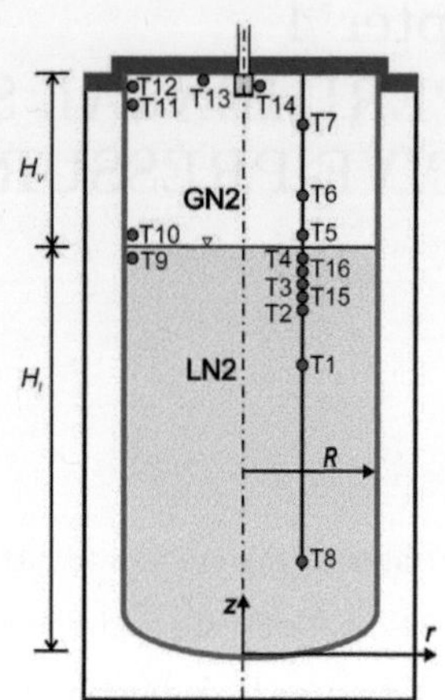

Fig. 4.2 Tank Setup 1: High pressure tank with the positions of the temperature sensors. Liquid level H_l at $z = 0.445$ m, vapor height $H_v = 0.205$ m.

Fig. 4.3 Tank Setup 2: High pressure tank with two additional temperature sensors (T15, T16) and raised liquid level H_l to $z = 0.455$ m, vapor height $H_v = 0.195$ m.

used a similar test setup for the investigation of sloshing of cryogenic liquids under normal gravity conditions.

The test facility used for this study is schematically depicted in Figure 4.1. It consists of a storage bottle for the pressurant gas, a mass flow controller, to ensure a constant pressurant gas mass flow, a heat exchanger, to control the pressurant gas temperature and a high pressure tank, partly filled with liquid nitrogen, in which the active-pressurization takes place. Valves are used to control the flow (V_{number}) and the pressure sensors are indicated as P_{number}.

The pressurant gas flows from the pressurant gas storage bottle, which is either a gaseous nitrogen or helium bottle, into a mass flow controller that ensures a constant pressurant gas mass flow. The pressurant gas is then fed through a heat exchanger to control the pressurant gas temperature. After passing the heat exchanger, the pressurant gas is either leaving through the bleed (if valve V_3 is open and V_2 is closed), in order to chill the lines down before the pressurization tests, or fed into the tank if V_3 is closed and V_2 open.

More detailed views of the high pressure tank are represented in Figures 4.2 and 4.3. The tank is a cylindrical test tank with a round shaped bottom and a tank radius of $R = 0.148$ m. The space between the outside casing and the inner container is evacuated and the related pressure is logged by the pressure sensor p_3 (see Figure 4.1). The inner casing is additionally insulated by 32 layers of multilayer insulation. The test tank has an internal volume of $43 \cdot 10^{-3}$ m^3 and is filled by two thirds with LN2. The tank wall is made of stainless steel with a wall thickness of $1.5 \cdot 10^{-3}$ m in the top third and below, the wall thickness is $2.0 \cdot 10^{-3}$ m. For the presented experiments, two different fill levels are used. In the Tank Setup 1 (Figure 4.2), the liquid-vapor interface is at $z = H_l = 0.445$ m and the height of the tank ullage is $H_v = 0.205$ m. In the Tank Setup 2 (Figure 4.3), the liquid-vapor interface is at $z = H_l = 0.455$ m ($\pm\ 1 \cdot 10^{-3}$ m) and the height of vapor phase is $H_v = 0.195$ m ($\pm\ 1 \cdot 10^{-3}$ m). The entering pressurant gas is distributed by the diffuser, which is a sintered filter (see Figure 4.4). The diffuser dimensions can be found in Figure A.1 in the Appendix. The used filter is sealed at the bottom so that the pressurant gas can only leave radially, in order to protect the liquid surface from a direct jet. With regard to the porosity of the filter can be stated that according to Dullien [35] for a random close pack of spheres, as for the filter, a porosity of ≈ 0.36 can be assumed. The pressure sensor p_1 logs the tank pressure and sensor p_2 the pressure of the gas before the injection into the tank. The maximum allowed tank pressure is 450 kPa.

The temperature inside the tank is logged at 14, respectively 16, pre-defined positions, depending on the test setup. The positions are marked with the gray dots in Figures 4.2 and 4.3. In Tank Setup 1 (Figure 4.2), the temperature sensors T1 to T8 are placed inside the tank on a glass fiber reinforced plastic retainer. The sensors T1, T2, T3 and T8 are placed in the liquid nitrogen at different heights and sensors T4 to T7 are mounted in the tank ullage. The positions of the sensors T4 and T3 is defined as just above and below the free surface respectively. The sensors T9 to T12 measure the tank wall temperatures and T13 the temperature at the inner side of the lid. The temperature sensor T14 is placed at a distance of $2 \cdot 10^{-3}$ m next to the diffuser in order to measure the temperature of the injected pressurant gas (see Figure 4.4).

For Tank Setup 2 (Figure 4.3), the liquid level lies between the sensors T4 and T5 and two additional sensors T15 and T16 are used in the liquid phase. The positions of the temperature sensors are summarized in Table A.1 in the Appendix. Additionally for both setups the sensor T17 measures the temperature of the pressurant gas in the feed line (see Figure 4.1).

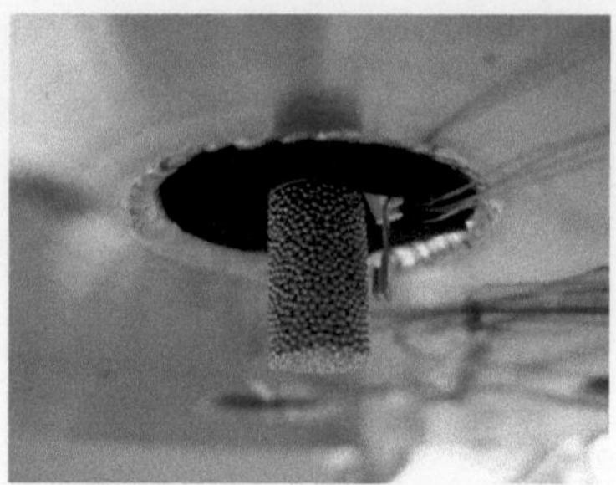

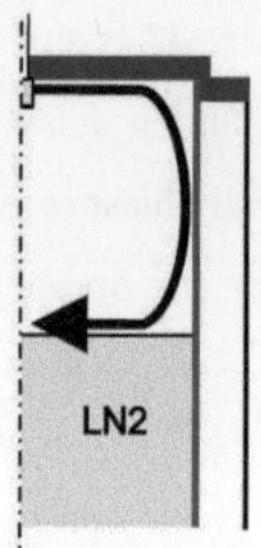

Fig. 4.4 Temperature sensor T14 placed next to the diffuser

Fig. 4.5 Schematical pressurant gas flow pattern.

4.2 Instrumentation

The instrumentation of the test facility for the mass flow, pressure and temperature measurement is described in the following sections. Specifications and the relating measurement errors are summarized in Table A.2 in the Appendix.

The pressurant gas mass flow rate is controlled by a mass flow controller (m1 in Figure 4.1, data see Table A.2 in the Appendix). It has a maximum mass flow rate of $8.62 \cdot 10^{-4}$ kg/s for air at 101.3 kPa and 273.15 K with an error of $\pm\, 13.9 \cdot 10^{-6}$ kg/s. The resulting mass flow rates for the experiments are $8.32 \cdot 10^{-4} \pm 13.2 \cdot 10^{-6}$ kg/s for GN2 and $1.62 \cdot 10^{-4} \pm 2.7 \cdot 10^{-6}$ kg/s for GHe (Fluidat [17]).

The pressure inside the tank is measured by the sensor p_1 (for data see Table A.2 in the Appendix) with an application range of 0 to 10^3 kPa and an accuracy of $\pm\, 7.4$ kPa. The pressure sensor p_2 inside the pressurization line (see Figure 4.1) is a Honeywell TJE 060/E663 03 TJA. This sensor has a maximum pressure of 500 kPa absolute and the accuracy is $\pm\, 3.2$ kPa. The pressure sensor p_3 (Penningvac PTR90) is used for the control of the vacuum pressure between the outside casing and the inner container of the tank. The vacuum pressure for the performed tests is in the range of $0.1 \cdot 10^{-6}$ to $0.3 \cdot 10^{-6}$ kPa. The accuracy of the pressure sensor is $\pm\, 0.175 \cdot 10^{-6}$ kPa. The data acquisition rate for the pressure sensors p_1 and p_2 is 10 Hz.

The temperature inside the tank is measured by 16 silicon diodes (for data see Table A.2 in the Appendix), designed for cryogenic data acquisition. The positions of the silicon diodes inside the tank are marked as T1 to T16 in Figure 4.3. The data of these sensors are logged at 3 Hz. The provider of the temperature sensors gives an accuracy of $\pm\, 0.5$ K. An internal evaluation showed that a higher accuracy of $\pm$ 0.1 K can be achieved. The emission of heat per sensor is $10 \cdot 10^{-6}$ W at 77 K and $5 \cdot 10^{-6}$ W at 300 K. The temperature sensor T17 in the pressurization line (see Figure 4.1) is a Pt100 temperature sensor. The accuracy from 73 K to 373 K is $\pm$ 2.1 K, the data is logged with a sample rate of 10 Hz. The maximum emission of heat is $50 \cdot 10^{-6}$ W at 373 K.

4.3 Test Procedure

This section describes the procedure for the active pressurization tests with GN2 and GHe as pressurant gases. In order to guarantee a single fluid system inside the tank, the empty tank is evacuated to 3.5 kPa before the first experiment. The valve V_3 and the storage bottle valve V_1 are then opened, the valve V_2 at the tank inlet is closed, and the feed lines are flushed with GN2 for approximately one minute. After that V_2 is opened, V_3 is closed and the tank is pressurized to above ambient pressure. An outlet port at the lid is then opened to allow a constant exhaust of GN2. With this GN2 outflow and the constant GN2 inflow from the storage bottle, the tank is filled with LN2 through an additional port in the lid. For refilling, the same procedure is used, except the evacuation phase.

In order to guarantee the same initial stratification conditions for each experiment run, the tank is overfilled with LN2 at a pre-set time several hours before the experiment start, and the tank outlet valve V_4 is partly opened so that the evaporated nitrogen can leave the tank. Due to the continuous outflow, the initial tank pressure p_0 for the experiments lies slightly over ambient pressure (see Table A.8 in the Appendix). When the liquid surface reaches the pre-defined position in the middle between the temperature sensors T3 and T4, the actual active-pressurization experiment is started: the bleed V_3 is then opened and the feed lines from the pressurant gas storage bottle to the branching are chilled down, respectively heated up. The tank outlet valve V_4 is then closed. After that, the inlet valve V_2 is opened simultaneously as V_3 is closed. The tank is now pressurized. When the final tank pressure is reached, the inflow is stopped by closing the tank inlet valve V_2 so that no mass can enter or leave the tank. Inside the tank, relaxation takes place. After a pre-set time, the tank outlet valve is opened again and the experiment is completed. During the whole experiment, the tank pressures and the temperatures are logged.

For this study, either GN2 or GHe is used as the pressurant gas. For the GN2 experiments, four different pressurant gas temperatures T_{pg} are chosen, measured at T17 (see Figure 4.1),

Table 4.1 Experimental matrix of the performed experiments with experiment name, applied pressurant gas, pressurant gas temperature T_{pg} and final tank pressure p_f. The experiment name consists of an abbreviation for the pressurant gas (N or He), the final tank pressure (e.g. 200 for 200 kPa) and an index for the pressurant gas temperature (e.g. c is $T_{pg} = 263$ K). T_{pg} is measured at temperature sensor T17.

Exp.	press. gas	T_{pg} [K]	p_f [kPa]
N200r	GN2	144	200
N300r	GN2	144	300
N400r	GN2	144	400
N200c	GN2	263	200
N300c	GN2	263	300
N400c	GN2	263	400
N200a	GN2	294	200
N300a	GN2	294	300
N400a	GN2	294	400
N200h	GN2	352	200
N300h	GN2	352	300
N400h	GN2	352	400
N300aH	GN2	294	300
He200c	GHe	263	200
He400c	GHe	263	400
He200h	GHe	352	200
He400h	GHe	352	400

to reach three different final tank pressures p_f and for the GHe pressurization two pressurant gas temperatures with two final tank pressures are selected (see Table 4.1).

4.4 Environmental Heat Transfer

For the determination of the heat flows from the ambient surroundings into the tank, a boil-off experiment was performed. Therefore, the test tank is partially filled with LN2 and the valves V_2 and V_3 were opened so that GN2 can leave the tank through the diffuser (see Figure 4.6). The amount of evaporated nitrogen is measured by placing the mass flow controller m2 after the valve V_2, which

determines the amount of GN2, leaving the tank (data for m2, see Table A.2 in the Appendix). Additionally, the temperature of the emergent gas is also measured with the sensor T17.

This experiment was run for 15 hours. However for the analysis presented here, only the data that was logged as the filling level corresponds to the filling level of Tank Setup 1 is used

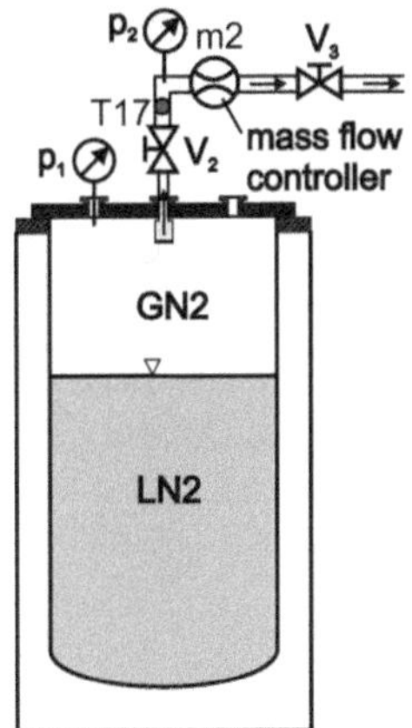
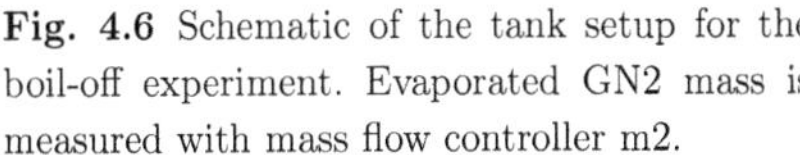

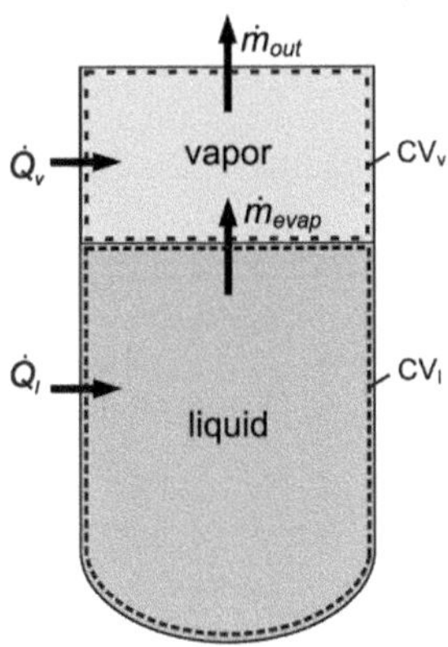

Fig. 4.6 Schematic of the tank setup for the boil-off experiment. Evaporated GN2 mass is measured with mass flow controller m2.

Fig. 4.7 Propellant tank with vapor and liquid phase, the referring control volumes and the heat and mass flow rates, considered for the boil-off analysis.

(Figure 4.2). It is assumed that the internal energy of the tank system stays constant over the considered time frame, as the tank pressure and the temperature profiles do not change significantly. On that account, the following form of Equation 2.34 for a closed, isobaric system can be applied

$$\dot{Q} = \dot{m}_{out}\, h_{out} \tag{4.1}$$

with $\dot{Q}$ as the net heat flow rate, entering the tank system and $\dot{m}_{out}$ the mass flow rate leaving the tank system with the specific enthalpy h_{out}.

The heat flow rates, required to evaporate that specific amount of liquid nitrogen can hence be calculated. The temperature of the leaving gas measured at T17 was 277.7 K and the tank pressure was 117 kPa, which results in a saturation temperature of the nitrogen of $T_{sat} = 78.6$ K. With the corresponding enthalpies $h_v(T_{sat}) = 78.09 \cdot 10^3$ J/kg and $h_l(T_{sat}) = -119.47 \cdot 10^3$ J/kg and $h_v(T_{T17}) = 287.93 \cdot 10^3$ J/kg, and the mass flow rate of $\dot{m}_{out} = 87.1 \cdot 10^{-6} \pm 4.5 \cdot 10^{-6}$ kg/s, the resultant heat flow rates are determined as

$$\dot{Q}_l = \dot{m}_{out} \left[h_v(T_{sat}) - h_l(T_{sat}) \right] = 17.2 \text{ W} \pm 0.9 \text{ W} \tag{4.2}$$

$$\dot{Q}_v = \dot{m}_{out} \left[h_v(T_{T17}) - h_v(T_{sat}) \right] = 18.3 \text{ W} \pm 1.0 \text{ W} \tag{4.3}$$

it therefore follows as result of the boil-off experiment that there is a total heat flow rate of $\dot{Q} = 35.5$ W ± 1.9 W from the environment into the test tank.

Chapter 5
NUMERICAL MODELING OF THE EXPERIMENTS

The performed experiments are analyzed numerically. Therefore the commercial CFD program Flow-3D is used. This chapter summarizes the relevant theoretical background of Flow-3D, the applied numerical model with the associated sensitivity analyses as well as the verification of the model.

5.1 Theoretical Background of Flow-3D

For numerical analysis of this study, Flow-3D version 10.0 is used. It is a general purpose computational fluid dynamics (CFD) code which numerically solves the equations of motion for fluids to determine transient, three-dimensional solutions to multi-scale, multi-physics flow problems. Fluid motion is described with non-linear, transient, second-order differential equations. A numerical solution of these equations involves approximating the various terms with algebraic expressions based on the consideration of control-volumes. The resulting equations are solved to get an approximate solution to the original problem. A numerical model starts with a computational mesh, or grid, which consists of a number of interconnected elements, or cells, subdividing the physical space into small volumes with several nodes that are associated with each such volume. These nodes are used to store values of the unknowns, such as pressure, temperature and velocity. Therefore, the mesh is effectively the numerical space, replacing the original, physical one and providing the means for defining the flow parameters at discrete locations, setting boundary conditions and for developing numerical approximations of the fluid motion equations. In Flow-3D, the flow domain is always subdivided into a grid of rectangular cells. The program can be used in several modes corresponding to different limiting cases of the general fluid equations. For instance, one mode is for compressible flows, and another is for purely incompressible flows. For incompressible flows, the fluid density and energy may be assumed constant and do not need to be computed. Additionally, there

are one-fluid and two-fluid modes. These operation modes correspond to different choices for
the governing equations of motion. In order to capture fluid interfaces in space and time, the
Volume of Fluid (VOF) method is employed in Flow-3D. Therefore, the fluid configuration
is defined in terms of the volume of fluid function. This function is equal to 1.0 at any point
inside fluid 1 and equal to 0.0 elsewhere. Averaged over one cell, it represents the volume of
fluid 1 per unit open volume, or fluid fraction. For two-fluid problems the complement of F,
which is 1-F, represents the volume fraction occupied by the fluid 2 [44].

In Flow-3D, phase change is modeled by

$$\hat{m}_{net} = rsize \sqrt{\frac{M}{2\pi \bar{R} \, T_{bdy}}} \left(p_l^{sat} - p_v\right) \tag{5.1}$$

where $\hat{m}_{net}$ is the net mass flux, $rsize$ is a net accommodation coefficient, M is the molar
mass of the vapor, $\bar{R}$ is the universal gas constant and T_{bdy} is the average liquid temperature along the free surface [44]. This equation is similar to Equation 2.54 of the interfacial
mass flux published by Carey [18], already introduced in Section 2.4. Both include an accommodation coefficient, which describes the rate of the phase change, but which are not
identically defined: Carey uses $\hat{\sigma}$ as accommodation coefficient and in Flow-3D, $rsize$ is defined as accommodation coefficient. As for the accommodation coefficient of Carey $\hat{\sigma}$, also
for the Flow-3D accommodation coefficient $rsize$, no recommended value is published for
cryogenic propellants. In Flow-3D, $rsize$ is defined as a multiplier on the phase change rate,
which has therefore a direct impact on the calculated phase change mass.

5.2 Flow-3D Model of the Test Tank

The experiment tank is modeled in a cylindrical coordinate system. The mesh is chosen as
quasi 2-D cylindrical mesh with an opening angle of 1° and the z-axis as axis of symmetry (see
Figure 5.1). The used fluid properties are taken from the NIST database [62] and are summarized in Tables 5.1 to 5.3. For the performed simulations, phase change is considered with the
saturation pressure $p_{sat} = 101.3$ kPa at $T_{sat} = 77.35$ K and $\Delta h_v = 199180$ J/kg and the expo-
nent in the relationship between saturation pressure and temperature is $tvexp = 0.0014$ 1/K.
The system is modeled using a two fluid model with a free surface under the consideration
of gravity, fluid to solid heat transfer, the first order approximation to the density transport
equation and the two fluid interface velocity slip. The pressurant gas is brought into the
tank by a mass source in vertical direction, placed at the position and with the dimensions

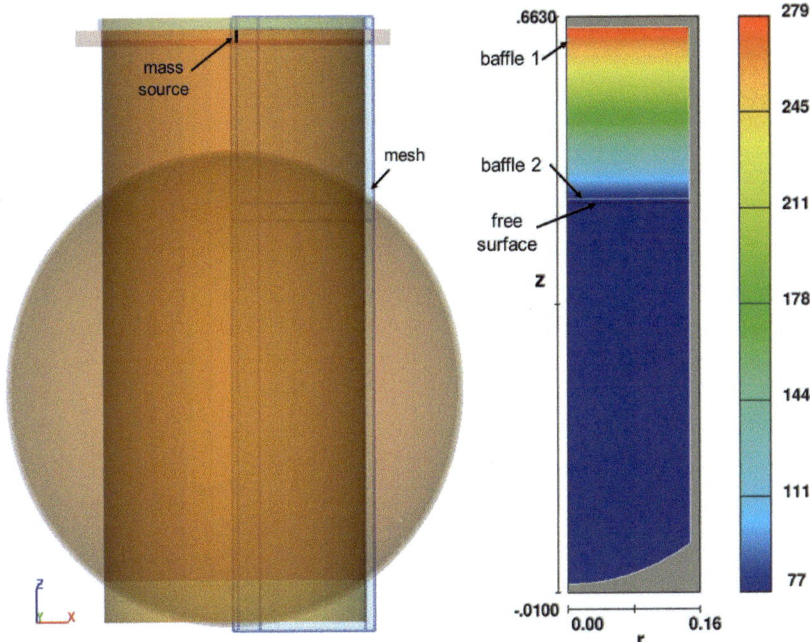

Fig. 5.1 Flow-3D model of the test tank with the mass source and 2-D view of the cylindrical mesh.

Fig. 5.2 Initial temperature contour of the fluids (in K) with the free surface ($z = 0.445$ m), the baffle over the free surface (baffle 2) and the baffle below the mass source (baffle 1).

of the diffuser (see Figure 5.1). The pressurant gas mass flow rate for GN2 is $2.31 \cdot 10^{-6}$ kg/s and $0.451 \cdot 10^{-6}$ kg/s for GHe with the corresponding pressurant gas temperatures T_{pg} summarized in Table 4.1. An additional baffle with a porosity of 0 is implemented right below the mass source to ensure the vertical flow in the first cell row next to the mass source (baffle 1 in Figure 5.2). If helium is used as pressurant gas, the non-condensable gas model is used. For both pressurant gases, the viscous laminar flow model is applied. An additional baffle has to be placed above the free surface ($z = 0.45$ m) with the porosity of 1 and no heat transfer properties (baffle 2 in Figure 5.2). This is necessary for the simulations due to the fact that the free surface gets destroyed otherwise by the pressurant gas flow. It is found however, that during the performed experiments the free surface is not affected by the pressurant gas flow. The lid of the experimental tank has a constant temperature of 280 K. The initial state for the simulations is: LN2 has a liquid temperature of 77 K and the vapor phase is thermally stratified, using the data from the initial temperature profile of the experiments (see temperature contour in Figure 5.2). The tank wall has the same initial temperature as the fluids. The data for the thermal conductivity of the tank walls is applied temperature dependent according to Barron [11] (curve for 304 stainless steel in Figure A.2 in the Appendix). Around the tank wall, a perfect insulation is modeled, which allows no heat transfer to or from the tank wall. This assumption is valid as the test tank is covered by 32 layers of multilayer insulation.

The initial tank pressure is taken from the experimental data and the implicit pressure solver GMRES is selected. As initial time step $1 \cdot 10^{-5}$ s is selected with a maximum time step of 0.001 s.

Table 5.1 Fluid properties for LN2 used in Flow-3D at 77.35 K and 101.3 kPa.

ρ_{LN2}	806.11 kg/m^3
μ_{LN2}	160.69·10^{-6} Pa s
$c_{v,LN2}$	1084.1 J/(kg K)
λ_{LN2}	0.14478 W/(m K)

Table 5.2 Fluid properties for GN2 used in Flow-3D at 77.35 K and 101.3 kPa.

$R_{s,GN2}$	296.8 J/(kg K)
ρ_{GN2}	4.6096 kg/m^3
μ_{GN2}	5.4436·10^{-6} Pa s

Table 5.3 Fluid properties for GHe used in Flow-3D at 77.35 K and 101.3 kPa.

$R_{s,GHe}$	2077 J/(kg K)
$c_{v,GHe}$	3115.9 J/(kg K)

5.3 Sensitivity Analysis

The cell size of the mesh and the Flow-3D accommodation coefficient *rsize* are determined as the main factors influencing the numerical results. A sensitivity analysis is therefore performed with variation of both factors. The mesh is chosen as quasi 2-D cylindrical mesh with with an opening angle of 1° in y direction and the z-axis as axis of symmetry (see Figure 5.1). In the sensitivity analysis, the following cell sizes are analyzed for the x and z direction: 0.002x0.002 m, 0.0025x0.0025 m, 0.003x0.003 m, 0.002x0.0025 m. The following values for the Flow-3D accommodation coefficient *rsize* are additionally evaluated: 0.1, 0.01, 0.001 and 0.0001.

Figure 5.3 depicts the results of the pressurization phase for the different mesh sizes and *rsize* parameters of the Flow-3D model in comparison to the experimental data of the N400a experiment. The configurations with the best accordance of the experimental and numerical data are the 0.0025x0.0025 m mesh with *rsize* values of 0.1 and 0.01 and the 0.002x0.0025 m mesh with a *rsize* value of 0.1. In the numerical calculations of the

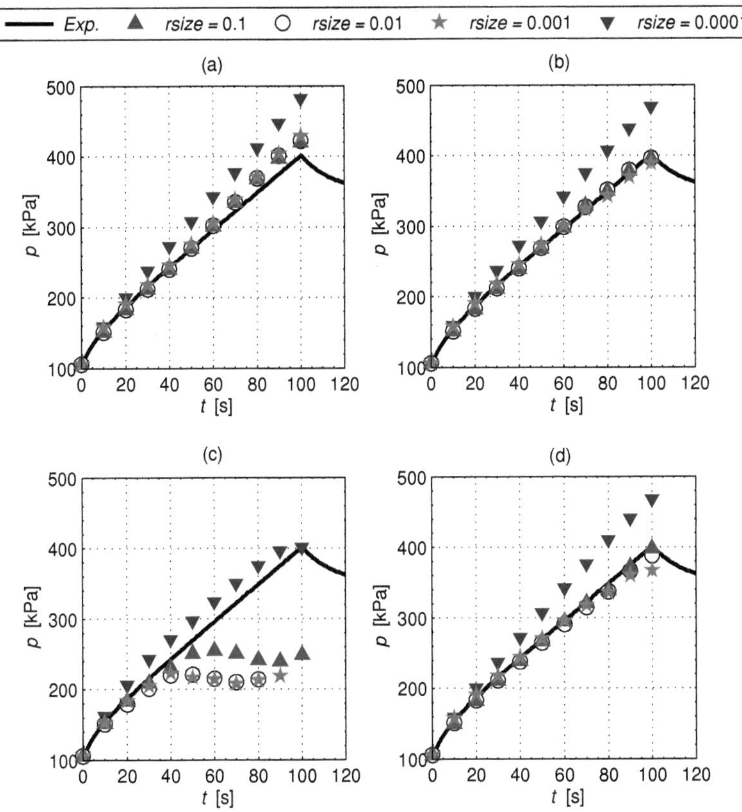

Fig. 5.3 Sensitivity analysis of the Flow-3D model: Numerical results for the pressure curve with different mesh sizes and Flow-3D accommodation coefficients *rsize* in comparison to the N400a experiment's pressure curve. The applied cell sizes for the mesh in x and z direction: (a) 0.002x0.002, (b) 0.0025x0.0025, (c) 0.003x0.003, (d) 0.002x0.0025.

remaining experiments it is found that the only mesh, with which all experiments can be simulated with meaningful results is the 0.002x0.0025 m mesh. During all performed Flow-3D simulations, the free surface dithers due to parasitic currents. Parasitic currents are unphysical currents, generated near the free surface by local variations in the surface tension forces [52]. A converging behavior of the cell size and/or the *rsize* value is not found. A mesh with a cell size smaller than 0.002 m is not possible as an error appears, which says that the number of particles at the mass source exceeds the maximum allowable number of particles. Please note that no tracer particles are used for this simulation but the error message relates to the mass flux over the cell boundary, which is too high. On that account, the 0.002x0.0025 m mesh is always applied for the numerical simulations of this study. For the simulations of all GN2 pressurized experiments, the *rsize* value of 0.1 is used. For the simulation of the GHe pressurized experiments, the value of *rsize* has to be decreased to 0.0001 to enable numerical results. All applied *rsize* values are summarized in Table 5.4.

5.4 Verification of the Numerical Model

Table 5.4 Applied Flow-3D accommodation coefficient *rsize* for the simulations of the performed experiments.

Exp.	rsize
N200r	0.1
N300r	0.1
N400r	0.1
N200c	0.1
N300c	0.1
N400c	0.1
N200a	0.1
N300a	0.1
N400a	0.1
N200h	0.1
N300h	0.1
N400h	0.1
N300aH	0.1
He200c	0.0001
He400c	0.0001
He200h	0.0001
He400h	0.0001

For the verification of the Flow-3D model, the pressure evolution of the active-pressurization phase is compared to the corresponding experimental data. The Flow-3D model, introduced in Section 5.2 with the 0.002x0.0025 m mesh and the *rsize* values of Table 5.4 are applied. Figure 5.4 depicts the results of the Flow-3D simulations for the active-pressurization phase with the final tank pressure of 400 kPa for the GN2 pressurized experiments in comparison to the experimental data. The curves with 400 kPa as final tank pressure also cover the experimental and numerical cases with 200 kPa and 300 kPa final tank pressure, as the pressure slope is identical, which will be shown in Figure 6.9. The numerical results for the N400h, N400a and N400c experiments in Figure 5.4 show very good agreement with the experimental data, which means that for these cases the numerical model is well suited. For the simulation of the N400r experiment, deviation appears. It can be seen in Figure 5.4 (d) that the numerical pressure curve increases almost linearly, the experimental pressure increase however, has a more curved evolution. This might be due to the fact that the phase change mass flow rate calculated by Flow-3D from pressurization start up to a tank pressure of about 300 kPa is higher than appearing in the experiment. Afterwards, Flow-3D probably underestimates the phase change mass flow rate, resulting in a faster pressure increase than in the experiment.

Figure 5.5 depicts the numerical and experimental results of the helium pressurized experiments He400h and He400c. For the simulation of these experiments *rsize* has to be decreased

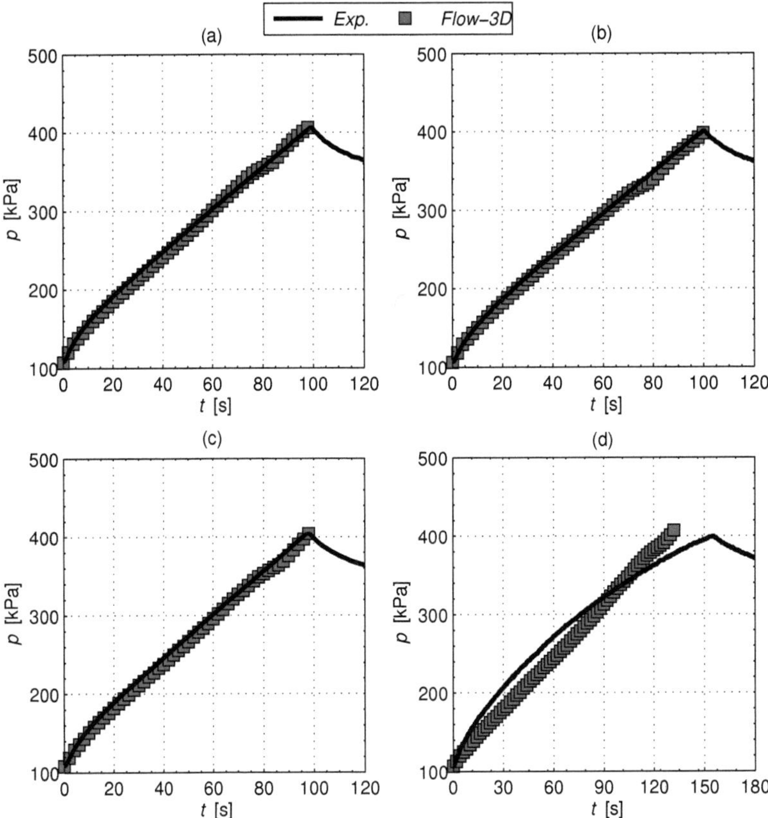

Fig. 5.4 Comparison of experimental pressure curve and results of the applied Flow-3D model for the (a) N400h (b) N400a (c) N400c and (d) N400r experiments.

to 0.0001 to get stable simulations. It can be seen that for these test cases, the discrepancy between the experimental and numerical data is quite big compared to the GN2 pressurized cases. This discrepancy is assumed to be due to the low value of *rsize* and the more complex numerical computation of the gas mixture with a non-condensable gas in the vapor phase.

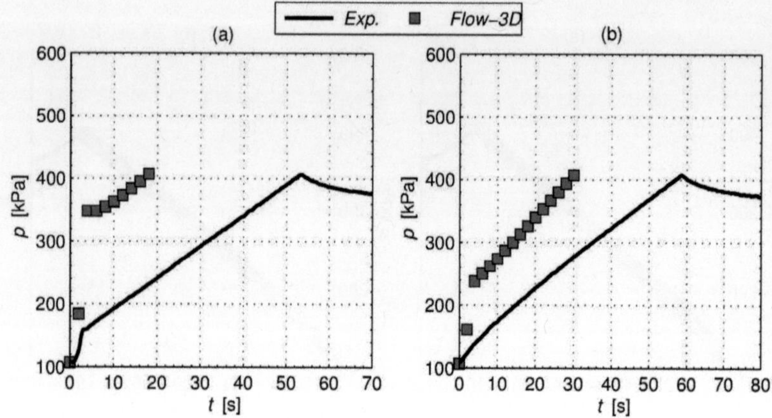

Fig. 5.5 Comparison of experimental pressure curve and results of the applied Flow-3D model for the (a) He400h and (b) He400c experiments.

Chapter 6
EXPERIMENTAL, THEORETICAL AND NUMERICAL RESULTS

This chapter will present the main results of the performed experiments together with the results from numerical simulations and theoretical approaches.

1. First, the data scaling, used for the nondimensional representation of the experimental results is introduced.

2. The experimental results for the pressure and temperature evolution during and after active-pressurization are then presented.

3. The evolution of the thermal stratification in the liquid and vapor phase during and after the pressurization is subsequently evaluated. The experimental data is compared with analytical heat transfer models in order to identify the dominating heat transfer mechanisms.

4. Moreover, the pressurant gas mass, required to reach the final pressure level, is analyzed with respect to the pressurant gas type and its inlet temperature. The experimental results are compared to results of the numerical simulations.

5. Furthermore, by means of analytical considerations, the phase change during and after pressurization is analyzed and the results are compared to the numerical results.

6. Along with this, an assessment of the heat flows during active-pressurization and relaxation is presented, which is based on experimental, analytical and numerical analyses.

7. Furthermore, a correlation is presented which allows an a priori estimation of the pressure rise during the pressurization phase.

8. Finally, the pressure drop after the pressurization end is analyzed. Therefore, the experimental data are compared to results of an analytical pressure drop model and the numerical simulations.

9. By means of one performed experiment and the relating numerical simulations, an overview of the main factors of active-pressurization is presented.

10. Ultimately, a correlation is presented for the determination of the required pressurant gas mass based on the JAKOB number and the thermal expansion FROUDE number.

6.1 Data Scaling

As established in Section 2.7, the experimental data in this study is presented in a nondimensional scaled form, in order to be able to compare the results to other experimental data and fluids. Please note that the scaling concept, presented hereafter is a geometrical scaling approach. From Equation 2.82, it follows that the nondimensional liquid temperature is defined as:

$$T_l^* = \frac{T_l - T_{ref}}{\Theta_l} = \frac{T_l - T_{ref}}{T_{sat,f} - T_{ref}} \tag{6.1}$$

The temperature $T_{sat,f}$ is the saturation temperature corresponding to the final tank pressure. This parameter is selected for the characteristic temperature difference as the final tank pressure is a key parameter for the active tank pressurization, which is restricted by propellant, structural and engine requirements. For the experiments with 200 kPa as the final pressure it is $T_{sat,f} = 83.6$ K, for 300 kPa final pressure $T_{sat,f} = 87.9$ K and for 400 kPa it is $T_{sat,f} = 91.2$ K. The reference temperature T_{ref} is the saturation temperature at norm pressure $p_{norm} = 101.3$ kPa, which is for nitrogen $T_{ref} = 77.35$ K. The nondimensional liquid temperature is scaled between 0 and 1: for $T_l = T_{ref}$, it is $T_l^* = 0$ and for $T_l = T_{sat,f}$, it is $T_l^* = 1$.

The nondimensional vapor temperature T_v^* is defined as

$$T_v^* = \frac{T_v - T_{ref}}{\Theta_v} = \frac{T_v - T_{ref}}{T_{pg,m} - T_{ref}} \tag{6.2}$$

consequently, the nondimensional wall temperature T_w^* is

$$T_w^* = \frac{T_w - T_{ref}}{\Theta_v} = \frac{T_w - T_{ref}}{T_{pg,m} - T_{ref}} \tag{6.3}$$

and the dimensionless pressurant gas temperatures T_{pg}^* is

$$T_{pg}^* = \frac{T_{pg} - T_{ref}}{\Theta_{pg}} = \frac{T_{pg} - T_{ref}}{T_{pg,m} - T_{ref}} \tag{6.4}$$

where $T_{pg,m}$ is the maximum pressurant gas temperature at the subsystem inlet. This parameter is selected, as it represents the maximal possible temperature in the vapor phase. For the presented experiments $T_{pg,m} = 352$ K. The nondimensional vapor, wall and pressurant gas temperatures are scaled between 0, which occurs if T_v^*, T_w^* or T_{pg}^* is equal the reference temperature T_{ref} and 1, if T_v^*, T_w^* or T_{pg}^* is equal to $T_{pg,m}$.

As the main focus of this study lies on the ullage of the propellant tank, the tank pressure is scaled by the characteristic thermodynamic pressure

$$p^* = \frac{p}{\rho_{ref,v} \, R_s \, T_{ref}} \tag{6.5}$$

where $\rho_{ref,v} = 4.61$ kg/m^3 is the reference density of GN2 at norm pressure and the specific gas constant of GN2 is $R_{s,GN2} = 296.8$ J/(kg K). The majority of the depicted pressure curves are smoothened for better presentation with a local regression using weighted linear least squares and a second order degree polynomial model. Table A.6 in the Appendix summarizes the applied coefficients.

The nondimensional pressurant gas mass m_{pg}^* is defined by the required pressurant gas mass m_{pg} over the vapor mass in the tank ullage before pressurization $m_{v,0}$. Therefore, the average initial vapor mass over all experiments with a liquid height of $H_l = 0.445$ m is used, which results in $m_{v,0} = 0.035$ kg. For the N300aH experiment, which has an increased liquid level, $m_{v,0} = 0.032$ kg.

$$m_{pg}^* = \frac{m_{pg}}{m_{v,0}} \tag{6.6}$$

Every other vapor mass is scaled in the same way:

$$m_v^* = \frac{m_v}{m_{v,0}} \tag{6.7}$$

The nondimensional time t^* is defined for this study as

$$t^* = t \, \frac{D_{t,l}}{A_\Gamma} \tag{6.8}$$

with $D_{t,l}$ as the thermal diffusion coefficient of LN2 at norm pressure $D_{t,l} = 8.798 \cdot 10^{-8}$ m^2/s and A_Γ as the area of the liquid surface $A_\Gamma = 0.0688$ m^2. These parameters are selected as they describe heat transfer by thermal diffusion in the liquid phase in dependency of the area of the free surface of the propellant tank, which is relevant for thermal stratification analyses.

6.2 Experimental Pressure and Temperature Evolution

For the analysis of the active-pressurization process, it is not only important to consider the initial tank pressure, but also the temperature distribution inside the tank, as it has a significant effect on the tank's thermodynamic behavior during the active-pressurization. For the introduction of the pressure and temperature evolution during and after pressurization, the experiment N300h is chosen (for details see Table A.8 in the Appendix). For this experiment, the Tank Setup 1 is used (see Figure 4.2).

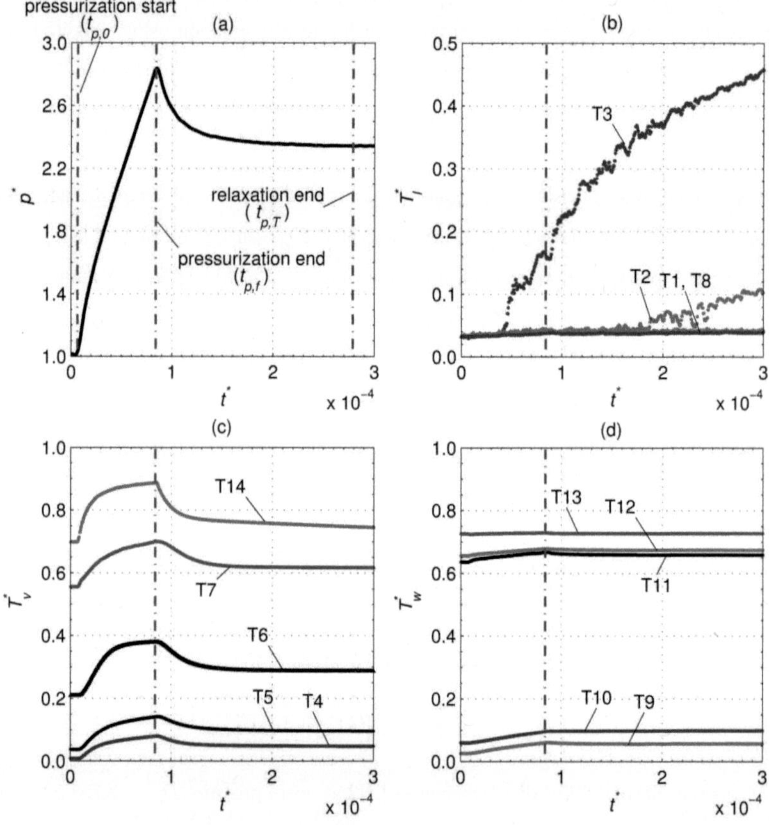

Fig. 6.1 (a) Tank pressure, (b) liquid temperatures, (c) vapor temperatures, (d) wall and lid temperatures during pressurization and relaxation of the N300h experiment (Tank Setup 1 see Figure 4.2, detailed data in Table A.8 in the Appendix). T14 is the pressurant gas temperature at the diffuser. Pressurization starts at $t_{p,0}$ ($t^* = 0.06 \cdot 10^{-4}$) and ends at $t_{p,f}$ ($t^* = 0.84 \cdot 10^{-4}$). Relaxation takes place until $t_{p,T}$ ($t^* = 2.79 \cdot 10^{-4}$).

The initial state for the experiment is as follows: the tank ullage is filled only with evaporated nitrogen, representing a two phase system with a single fluid. The lid has a constant outer temperature of 280 K and, due to its construction, a nearly constant inner lid temperature of 278 K. The free surface is located in the middle of T3 and T4. The average initial tank pressure for all experiments is 106 kPa and the temperature of the free surface is always considered to be the saturation temperature of the current tank pressure (Baehr and Stephan [9]), e.g. 77.7 K for 106 kPa and 87.9 K for 300 kPa.

During the experiments, the tank pressure and the temperatures in the liquid, the vapor and at the tank wall, are logged. Figure 6.1 depicts the pressure and temperature evolution of the pressurization experiment N300h. The nondimensional liquid temperature is determined with $T_{sat,f} = 87.9$ K. For this experiment, GN2 is used as the pressurant gas with an inlet temperature of 352 K, measured at T17, which is also the maximum possible inlet temperature of the test setup used for this study. The final tank pressure after pressurization is 300 kPa.

Figure 6.1 (a) shows the tank pressure during and after the pressurization process. The tank pressure increases almost linearly from $t_{p,0}$, where the pressurization starts, up to the pressurization end at $t_{p,f}$. After the pressurization end, when the pressurant gas inflow is stopped, the pressure decreases instantaneously and reaches a minimum pressure at $t_{p,T}$ with an asymptotical evolution. This phase, called relaxation, is mainly caused by condensation and will be analyzed in Section 6.8. Section 6.4 will look more detailed into the pressure rise of the different experiments.

Figure 6.1 (b) depicts the evolution of the liquid temperatures. In the considered time frame, only the two uppermost temperature sensors T2 and T3 detect a change in temperature, whereas the bulk temperature (sensors T1 and T8) remains constant. The topmost sensor T3 traces the main temperature increase after the end of pressurization and the liquid temperature at T2 rises even later and with a weaker slope. This observation suggests that heat transfer through the liquid is driven by conduction, caused by heat input over the free surface. This will be analyzed in Section 6.3.

The temperature evolution of the vapor phase is depicted in Figure 6.1 (c). All temperature sensors show an increase in vapor temperature during the active pressurization and an asymptotic decreasing behavior after the pressurization end, which is similar to the pressure evolution. The temperature sensor T14, placed directly next to the diffuser, shows the evolution of the pressurant gas inlet temperature during the pressurization process. At T17 (see Figure 4.1), the pressurant gas has a constant temperature of 352 K due to the heat exchanger. However, the temperature at T14 is not constant over time since the connecting pipe between the tank inlet valve V_2 and the diffuser cannot be pre-conditioned before the pressurization because of the lines' setup. Therefore, the line and the diffuser have to adapt to the gas temperature during the pressurization process. Figure 6.1 (c) also shows that the hot pressurant gas increases the temperature of the vapor with decreasing impact from the lid downwards. A closer look at the evolution of the vapor temperature will be presented in Section 6.3.

Figure 6.1 (d) shows the evolution of the wall temperatures (T9 to T12) and the temperature at the inner side of the lid (T13). The wall temperature is also affected by the pressurization process, but much less than the vapor temperature due to the slow reaction of the wall material. It can be seen that the lid temperature T13 is not changing over time. Unlike the vapor phase decreases the temperature of the wall only little after the end of the pressurization. The tank wall is affected by heat flows, which will be evaluated in Section 6.6. The temperature at the inner side of the lid changes only slightly over the entire experiment.

Figure A.3 in the Appendix shows the pressure and temperature evolution of the N300h experiment from the pressurization start until the experiment was stopped. This includes also the phase after relaxation end which is given for information only as it is not part of this analysis.

6.3 Evolution of Thermal Stratification

Based on the experimental pressure and temperature evolutions, the development of the thermal stratification in the liquid and vapor phase during and after the active-pressurization can be evaluated. The evolution of the stratification of one selected experiment is subsequently analyzed and compared to results of analytical heat transfer models in order to identify the dominating heat transfer mechanisms.

For this analysis, Tank Setup 2 (as depicted in Figure 4.3) is used. In order to be able to better monitor the thermal boundary layer in the liquid, two additional temperature sensors T15 and T16 are attached to the temperature measurement retainer and the position of the liquid level is raised by 0.01 m to $H_l = 0.455$ m. Consequently, the vapor height is $H_v = 0.195$ m. The positions of all temperature sensors are summarized in Table A.1 in the

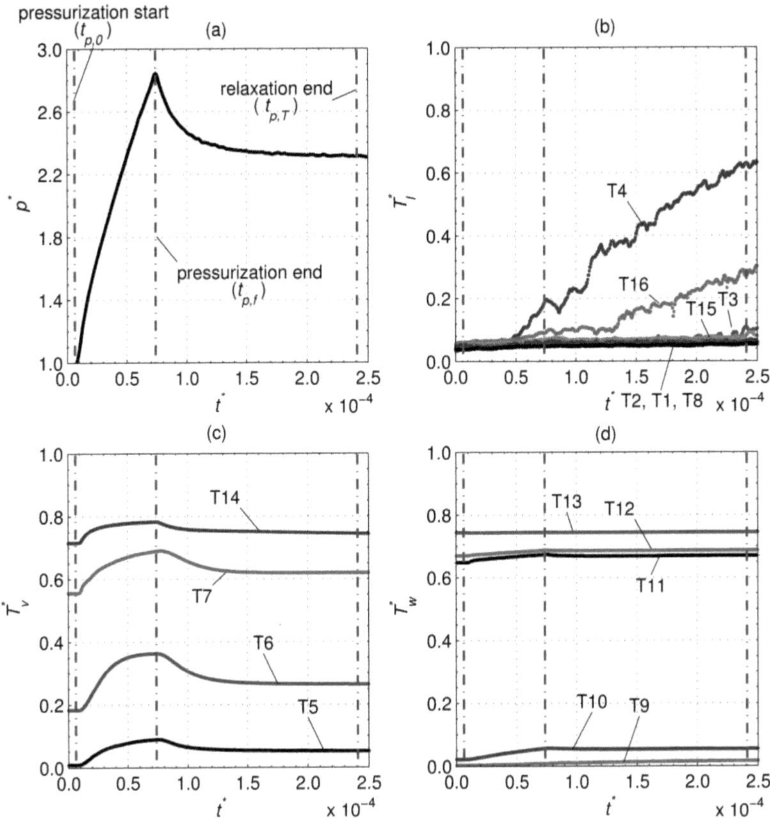

Fig. 6.2 (a) Tank pressure, (b) liquid temperatures, (c) vapor temperatures, (d) wall and lid temperatures during pressurization and relaxation of the experiment N300aH (Tank Setup 2 see Figure 4.3, detailed data in Table A.8 the Appendix). T14 is the pressurant gas temperature at the diffuser. Pressurization starts at $t_{p,0}$ ($t^* = 0.06 \cdot 10^{-4}$) and ends at $t_{p,f}$ ($t^* = 0.74 \cdot 10^{-4}$). Relaxation takes place until $t_{p,T}$ ($t^* = 2.41 \cdot 10^{-4}$).

Appendix. The times before pressurization start $(t_{p,0})$, at the end of pressurization $(t_{p,f})$ and after relaxation $(t_{p,T})$ are selected as the characteristic times (see Figure 6.2 (a)).

In Figure 6.2, the pressure and temperature evolution for the N300aH experiment are depicted. During the N300aH experiment, the tank is pressurized up to 300 kPa with a pressurant gas temperature of 294 K (for more details, see Table A.8 in the Appendix). The nondimensional liquid temperature is determined with $T_{sat,f} = 87.9$ K. The wall temperature T9 is made nondimensional with Equation 6.3 for comparison purpose in Figure 6.1 (d), although the sensor is placed in the liquid phase.

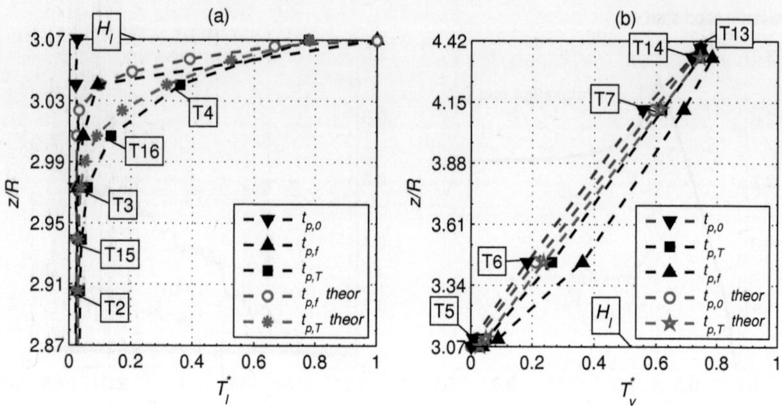

Fig. 6.3 (a) Liquid and (b) vapor temperature profiles of the N300aH experiment before pressure ramping ($t_{p,0}$), at pressurization end ($t_{p,f}$) and after relaxation ($t_{p,T}$). Theoretical liquid temperature profiles with Equation 6.9. Theoretical vapor temperature profiles with Equation 6.11. Saturation temperature at free surface ($z/R = 3.07$) is calculated; dashed lines are only for better visualization. All data can be found in Tables A.11 and A.7 in the Appendix.

Figure 6.3 shows for the selected times the profile of the vertical thermal stratification in the liquid and the vapor phases (all data can be found in Table A.11 in the Appendix). The dashed lines are only applied for better visualization and do not represent the actual temperature development between the data points. The free surface is at $z/R = 3.07$. In Figure 6.3 (a), only the temperature sensors T4 to T2 are depicted for better display, as the sensors T1 and T8 have the same temperature as T2 for all presented time steps. The temperature at the free surface is assumed to be the saturation temperature of the corresponding tank pressure (according to Baehr and Stephan [9]) and is calculated using the NIST database [62]: for $t_{p,0}$ it is 77.3 K, for $t_{p,f}$ 87.9 K and for $t_{p,T}$ it is 85.6 K.

For the liquid phase (Figure 6.3 (a)), no thermal stratification appears before the pressurization, due to the preceding propellant boil-off. Right after pressurization end, at $t_{p,f}$, a very sharp gradient in the temperature stratification emerges and the thermal boundary layer ranges from the free surface to approximately T3 ($z/R = 2.973$). After relaxation, at $t_{p,T}$, the thickness of the thermal boundary layer is increased to approximately T15 ($z/R = 2.939$) and the temperature gradient is weaker. Figure 6.4 is a schematic of the evolution of the thermal gradients at the three characteristic times: $t_{p,0}$ with no stratification, $t_{p,f}$ with sharp gradient and thermal boundary layer with the thickness $\delta_{T,f}$, and $t_{p,T}$ with the decreased temperature gradient and the thicker thermal boundary layer $\delta_{T,T}$.

Based on this experimental data, it is possible to apply analytic heat transfer models in order to predict the evolution of the temperature profile over time in the liquid phase. This approach, already presented by Ludwig et al. [66], is the approximation of transient heat transfer in large media with constant initial temperature (Equation 2.42). For this experiment, the initial temperature is the liquid temperature T_l and the boundary condition is that the surface temperature is suddenly changed and maintained at the new saturation temperature

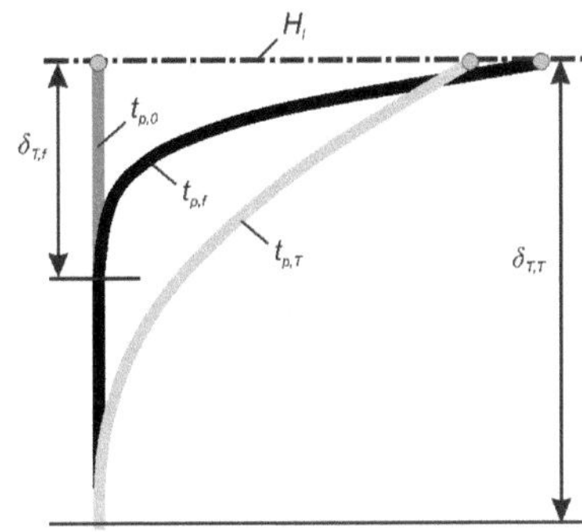

Fig. 6.4 Schematic of the evolution of the vertical temperature profiles and thickness of the thermal boundary layers in the uppermost part of the liquid from pressurization start $(t_{p,0})$ to pressurization end $(t_{p,f})$ and after pressurization end $(t_{p,T})$.

$T = T_{sat}$. Equation 6.9 is used for the prediction of the theoretical liquid temperature distribution at $t_{p,f}$ and $t_{p,T}$.

$$T(z) - T_l = (T_{sat} - T_l)\,\mathrm{erfc}\left[\frac{H_l - z}{2\sqrt{D_{t,l}\,t}}\right] \tag{6.9}$$

The time t is set to zero at the beginning of the pressurization and H_l is the height of the liquid phase (see Figure 4.3). Figure 6.3 (a) shows the comparison of the experimental data to the theoretical thermal stratification for $t_{p,f}$ with the thermal diffusion coefficient $D_{t,l} = 7.74 \cdot 10^{-8}$ m²/s and for $t_{p,T}$ with $D_{t,l} = 7.98 \cdot 10^{-8}$ m²/s. The thermal diffusion coefficients refer to the saturation temperature of the respective tank pressures. The theoretical approach used fits quite well for the time $t_{p,f}$ but underestimates the thickness of the thermal boundary layer for $t_{p,T}$, which may be due to the fact that the heat input from the wall and the latent heat set free due to condensation are disregarded in the theory. It can be summarized however, that the following similarity relationship between the pressurization duration and the thickness of the thermal boundary layer applies:

$$H_l - z \equiv 2\sqrt{D_{t,l}\,t} \tag{6.10}$$

It can be stated that for the pressurization and relaxation phase, the thermal boundary layer thickness in the liquid phase increases due to the change in saturation temperature at the free surface in the ratio of the root of the pressurization time.

Figure 6.3 (b) depicts the evolution of the thermal stratification in the tank ullage of the N300aH experiment from the free surface ($z/R = 3.07$) up to the inner side of the lid ($z/R = 4.39$). The corresponding data can be found in Table A.7 in the Appendix. At $t_{p,0}$, the temperature increases linearly between the saturation temperature at the free surface and the lid temperature. At $t_{p,f}$, the temperature sensor T14, next to the diffuser has the highest temperature and confirms that the temperature of the pressurant gas has a significant influence on the thermal stratification in the vapor phase. That strong thermal gradient decreases and after relaxation, an almost linear temperature gradient reappears. At $t_{p,T}$, the upmost layer of the ullage is still heated up from the warm pressurant gas and defines the gradient of the thermal stratification, which has below T14 again a nearly linear distribution to the temperature of the free surface.

For the ullage stratification, a theoretical approach is used in order to understand the behavior of the thermal stratification. For the thermal stratification at $t_{p,0}$ and $t_{p,T}$, the theoretical heat transfer model is the steady-state heat conduction in a flat plate (as introduced in Equation 2.44). The vapor temperatures are therefore calculated by Equation 6.11, where H_v is the height of the vapor phase.

$$T(z) = T_h + (T_{sat} - T_h)\frac{z_h - z}{H_v} \tag{6.11}$$

The temperature T_h represents the upper boundary condition. For the performed experiments, T_h at the time $t_{p,0}$ is defined as the temperature of the sensor T13 and at $t_{p,T}$ the temperature of the sensor T14. The height z_h is accordingly z_{13} and z_{14} at $t_{p,0}$ and $t_{p,T}$ respectively. The lower boundary condition is the saturation temperature T_{sat}.

Figure 6.3 (b) compares the results of the theory with the experimental data. For $t_{p,0}$, it can be seen that the theory overestimates the temperature of the sensors T6 ($z/R = 3.446$) and T7 ($z/R = 4.122$). This might be due to the fact that the theoretical approach only considers the lid as a heat source and disregards the tank walls.

For the time at pressurization end, no theoretical approach is presented due to the fact that the temperature profile is highly dependent on the pressurant gas temperature and can therefore not be predicted with analytical heat transfer models. Nevertheless, after relaxation, the theoretical approach of heat conduction can be applied again. The accordance between the experimental data and the theoretical results is very good. This is probably due to the fact that at this time, the tank walls and the ullage has enough time to adapt its temperatures.

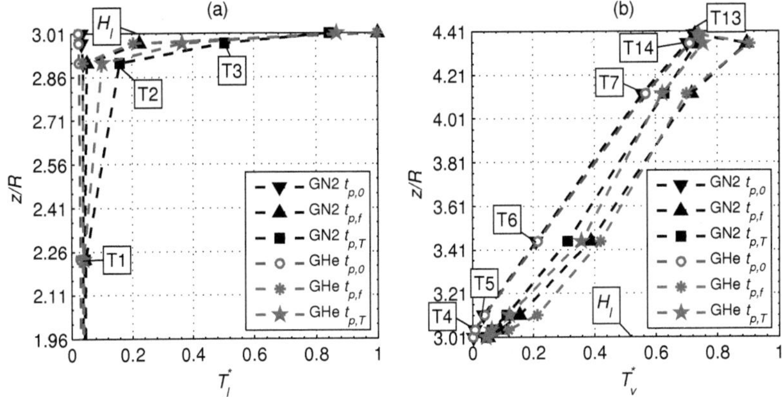

Fig. 6.5 (a) Liquid and (b) vapor temperature profiles of the N400h experiment (black markers) and the He400h experiment (gray markers) before pressure ramping $(t_{p,0})$, at pressurization end $(t_{p,f})$ and after relaxation $(t_{p,T})$. Saturation temperature at free surface $(z/R = 3.01)$ is calculated; dashed lines are only for better visualization. All data can be found in Table A.12 in the Appendix.

The thermophysical properties of gaseous helium and gaseous nitrogen differ substantially (see Table A.3 in the Appendix) and helium cannot condensate into the liquid nitrogen. On that account, the thermal stratification profiles of two relating experiments, one pressurized with GN2 and one with GHe, are compared. The experiments N400h and He400h are selected, as they both apply the highest pressurant gas temperature of 352 K and the highest final tank pressure of 400 kPa (for details see Table A.8 in the Appendix). For these experiments, the Tank Setup 1 is used (see Figure 4.2), with the free surface at $z/R = 3.01$, between the temperature sensors T3 and T4.

In Figure 6.5, the stratification data of the N400h experiment are depicted with the black markers and those for the He400h experiment with the gray markers (all data can be found in Table A.12 in the Appendix). Again, the dashed lines in this figure are only implemented for better visualization and do not represent the actual temperature development between the data points. In Figure 6.5 (a), only the temperature sensors T3 to T1 are depicted for better display, as the sensor T8 has the same temperature as T1 for all presented time steps. The temperature at the free surface is assumed to be the saturation temperature of the corresponding tank pressure (according to Baehr and Stephan [9]) and is calculated using the NIST database [62].

Figure 6.5 (a) shows the comparison of the liquid stratification for the GN2 pressurized N400h experiment and the helium pressurized He400h experiment for the pressurization start $(t_{p,0})$, the end of pressurization $(t_{p,f})$ and after relaxation $(t_{p,T})$. At pressurization end, almost no difference can be seen between the helium and the nitrogen pressurized experiment. However, at relaxation end, the temperature sensors T2 and T3 detected a higher temperature for the N400h experiment than for the He400h experiment. This might be on the one hand due to the fact that the helium prevents strong condensation during the relaxation and as a result, the latent heat of evaporation, set free during condensation in the N400h experiment increases the temperature in the thermal boundary layer. On the other hand, for the N400h experiment, the experimental time until relaxation end is considerably longer than for the He400h experiment (see Table A.8 in the Appendix). Due to that fact, the thermal boundary layer of the N400h experiment is thicker than for the helium experiment.

In the vapor phase (Figure 6.5 (b)), the thermal stratification at the pressurization end $(t_{p,f})$, shows a difference between the helium and nitrogen pressurized experiments for the sensors T4, T5 and T6. This is assumed to be due to the local distribution of helium and gaseous nitrogen in the ullage for the He400h experiment. For all three sensors, the helium pressurized experiment shows higher temperatures. After relaxation at $t_{p,T}$, only the sensor T6 still has a little higher temperature than the relating nitrogen pressurized experiment. It can be stated that for the helium pressurized experiments, a higher average vapor temperature appears during and after pressurization than for the GH2 pressurized experiments.

6.4 Required Pressurant Gas Mass

Ring [80] stated, that for intercontinental ballistic missile two-stages vehicles the reduction of the second-stage's weight is approximately ten times as important as that of the first-stage. This statement is also valid for launchers. On that account, high effort has to be made to design minimum-weight cryogenic upper stages for launchers. One objective of this study is to investigate the required pressurant gas mass for a defined tank pressure rise. On that account, the results of the experiments are presented in this section and subsequently compared to numerical results.

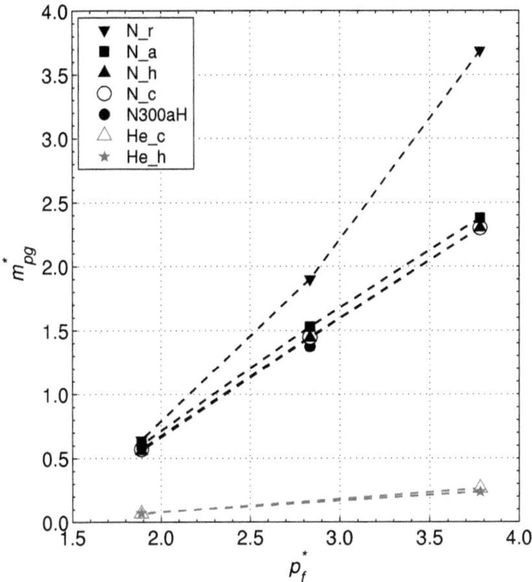

Fig. 6.6 Nondimensional required pressurant gas mass m_{pg}^* for all performed experiments over the nondimensional final tank pressure p_f^*. The dashed lines are only for better visualization, the errors lie within the size of the marker and the corresponding data can be found in Tables A.8 and A.9 in the Appendix.

6.4.1 Experimental Results

For each active-pressurization experiment, the amount of pressurant gas needed to increase the tank pressure between pressurization start and pressurization end is determined. The required pressurant gas mass is calculated from the pressurization time t_{press} ($t_{press} = t_{p,f} - t_{p,0}$) and the pressurant gas mass flow rate $\dot{m}_{pg}$, which is controlled by the mass flow controller (see Section 4.2).

$$m_{pg} = t_{press}\, \dot{m}_{pg} \tag{6.12}$$

The pressurant gas mass flow is kept constant at the maximum feasible mass flow of $8.62 \cdot 10^{-4}$ kg/s for air at 101.3 kPa and 273.15 K with an error of $\pm$ $13.9 \cdot 10^{-6}$ kg/s. The resulting mass flow for nitrogen is $8.322 \cdot 10^{-4} \pm 13.2 \cdot 10^{-6}$ kg/s and for helium $1.624 \cdot 10^{-4} \pm 2.7 \cdot 10^{-6}$ kg/s (Fluidat [17]). This corresponds to a maximum error of 5 % for m_{pg}. The pressurant gas inlet temperature is controlled during the experiments at temperature sensor

T17 (see Figure 4.1). The error for T_{pg} is $\pm$ 0.3 K, except for T_{pg} = 144 K, which has an error of $\pm$ 1.5 K due to the applied heat exchanger. All data, mentioned in this section is summarized in Tables A.8 and A.9 in the Appendix. The corresponding equations for the scaling were already presented in Section 6.1.

Figure 6.6 depicts the results of the dimensionless pressurant gas mass m_{pg}^* over the nondimensional final tank pressure p_f^* for all 17 performed experiments. It can be seen that the higher the final tank pressure, the more pressurant gas mass is required. This is due to the fact that a higher final tank pressure needs a longer pressurization time. This is valid for both GN2 and the GHe pressurization. Figure 6.9 compares the N200c and the N400c experiments, which have the same pressurant gas temperature but different final tank pressures. It can be seen that the pressure curves increase nearly identical for the pressurization phase. For the He200c and the He400c experiments, both are likewise pressurized with the same GHe temperature, also nearly identical pressurization curves are depicted.

Figure 6.6 shows that a smaller ullage volume results in a decrease in the required pressurant gas mass, as less volume has to be filled up: the N300aH experiment (black dot) has 5 % less ullage volume than the N300a experiment and requires 18 % less pressurant gas mass (for details see Table A.9 in the Appendix).

For all helium experiments, it is also seen that distinctly less pressurant gas mass is required than for the corresponding GN2 experiments. This is due to the following facts: helium has a much lower density than nitrogen, which means that less GHe is needed to fill the tank ullage (e.g. vapor density for helium at p = 106 kPa and a mean ullage temperature of T_v = 144 K is $\rho_{v,GHe}$ = 0.354 kg/m^3, and for nitrogen it is $\rho_{v,GN2}$ = 2.50 kg/m^3 [62]). Additionally, helium is a non-condensable gas and therefore cannot undergo phase change with the liquid nitrogen. Furthermore, the specific heat capacity c_p of gaseous helium is about 4.5 times as high as that of gaseous nitrogen (e.g. $c_{p,GHe}$ = 5.196·10^3 J/(kg K) and $c_{p,GN2}$ = 1.124·10^3 J/(kg K) at p = 101.3 kPa and T = 77.35 K [62]), resulting in an increased heat input into the vapor phase during the pressurization period, compared to the GN2 pressurization and therefore a faster pressure increase.

The required pressurant gas mass has been analyzed with respect to the different pressurant gas temperatures (the corresponding data can be found in Table A.9 in the Appendix). The highest required pressurant gas mass has the N400r experiment at T_{pg}^* = 0.24 (lowest pressurant gas temperature of 144 K). For the N400h experiment, the applied pressurant gas

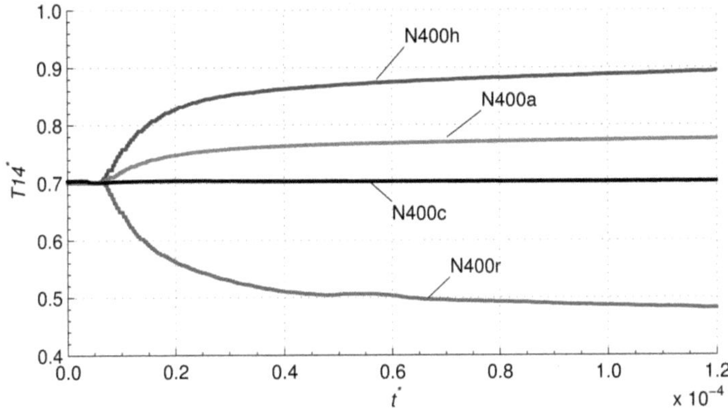

Fig. 6.7 Comparison of the evolution of the nondimensional pressurant gas inlet temperature T14* during pressurization for the N400h, N400a, N400c and N400r experiments (details see Table A.8 in the Appendix).

temperature is the highest pressurant gas temperature $T_{pg}^* = 1.00$ ($T_{pg} = 352$ K) and the required pressurant gas mass is about 38 % less than that required by the N400r experiment. For the N300r and N300h experiments, the difference in required pressurant gas mass is 24 % and for the N200r and N200h experiments it is approximately 13 %. The N400a, N300a and N200a experiments ($T_{pg}^* = 0.79$, $T_{pg} = 294$ K) fit well in the decreasing course of required pressurant gas mass between the experiments with $T_{pg}^* = 0.24$ and $T_{pg}^* = 1.00$.

However, the experiments with $T_{pg}^* = 0.68$ ($T_{pg} = 263$ K) require less pressurant gas mass than expected from the analysis of the other pressurant gas temperatures. For all five experiments, the pressurant gas mass is almost the same as for the experiments with a dimensionless pressurant gas temperature of $T_{pg}^* = 1.00$. This is due to the fact that for these experiments, the temperature of the pressurant gas injected at the diffuser fits very well to the ullage temperature at the height of the diffuser before pressurization. Hence, the diffuser and the pipe between the tank inlet valve and the diffuser did not need to be warmed up or cooled down as much as for the other pressurant gas inlet temperatures. Figure 6.7 compares the temperature evolution, measured at the sensor T14 for the GN2 pressurized experiments with a final tank pressure of 400 kPa. It can be seen, that for the N400c experiment, the nondimensional temperature T14* changes only slightly over time, whereas for the other depicted experiments, T14* changes strongly over time. This behavior is also valid for the GHe pressurization experiments.

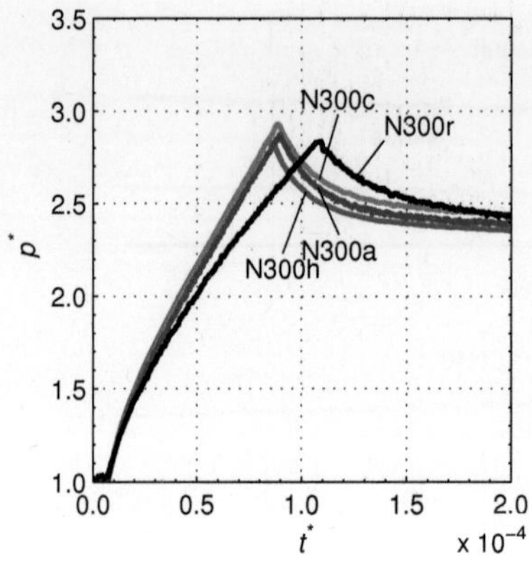

Fig. 6.8 Tank pressure evolution during pressurization and relaxation for the GN2 pressurization up to 300 kPa ($p_f^* = 2.84$) with the four different pressurant gas temperatures.

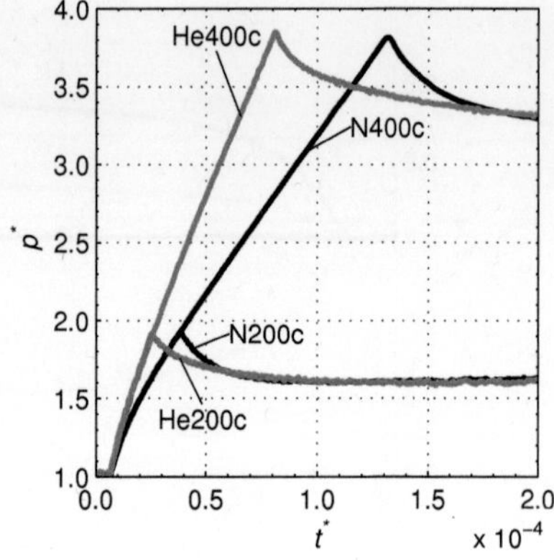

Fig. 6.9 Comparison of the tank pressure evolution during pressurization and relaxation for the GN2 and GHe pressurization up to 200 kPa and 400 kPa with the pressurant gas temperature of 263 K.

As the required pressurant gas mass is calculated from the needed pressurization time, four tank pressure rises over time are compared: Figure 6.8 depicts the pressure evolution of the N300h, N300a, N300c and N300r experiments, which are all pressurized with GN2 up to 300 kPa ($p_f^* = 2.84$). It can be seen that experiment N300r, with the lowest pressurant gas temperature, had the weakest pressure rise and therefore took the longest to reach 300 kPa. This points up why it needed the most pressurant gas mass of all N300 experiments. It can also be seen that the slope for the N300c experiment is almost identical to the N300h experiment, which explains the similar required pressurant gas masses. Experiment N300a has a little lower pressure curve and therefore requires more pressurant gas than the N300h and N300c experiments but less than the N300r experiment. Please note that not all curves have their peak at the same final pressure as the pressurization had to be stopped by manually closing the valve V_2. Nevertheless, all data used for this analysis considers only the pressurization time until the final pressure level of 300 kPa is reached ($p^* = 2.84$).

Figure 6.9 compares the pressure curves of two helium pressurized experiments with the corresponding GN2 experiments. It can be see that for both GHe pressurizations, the pressure increase is steeper than for the GN2 pressurizations. The faster pressurization is caused by the facts, presented earlier. This fast pressurization also contributes to the fact that all helium pressurizations require much less pressurant gas than the corresponding GN2 experiments, as Figure 6.6 had previously shown. As already mentioned, it can also be seen that for both GN2 and both GHe experiments, the pressure curves increase identically for the pressurization phase.

6.4.2 Numerical Results

Table 6.1 Experimental and numerical required pressurant gas masses of the active-pressurization phase with the relating errors. Errors are given of the Flow-3D results relative to the experimental masses. Please note that all values are rounded.

	Exp. [kg]	*Flow-3D* [kg]	*error* [%]	*abs. error* [kg]
N200c	0.0201	0.0216	7.9	0.0016
N300c	0.0507	0.0516	1.8	0.0009
N400c	0.0806	0.0807	0.2	0.0002
N200a	0.0212	0.0216	2.0	0.0004
N300a	0.0535	0.0516	-3.6	-0.0019
N400a	0.0833	0.0849	1.9	0.0016
N200h	0.0196	0.0208	5.9	0.0012
N300h	0.0505	0.0499	-1.2	-0.0006
N400h	0.0806	0.0799	-0.8	-0.0007
N200r	0.0225	0.0333	48.1	0.0108
N300r	0.0665	0.0707	6.4	0.0042
N400r	0.1292	0.1082	-16.2	-0.0209
N300aH	0.0439	0.0416	-5.1	-0.0022
He200c	0.0024	0.0008	-65.8	-0.0016
He400c	0.0093	0.0047	-49.3	-0.0046
He200h	0.0025	0.0005	-80.1	-0.0020
He400h	0.0084	0.0029	-65.0	-0.0054

Table 6.1 compares the required pressurant gas masses of the experiments and the Flow-3D simulations and gives the corresponding errors. The pressurant gas masses of the Flow-3D simulations are calculated with the results of the pressurization times, summarized in Table A.8 in the Appendix, multiplied by the pressurant gas mass flow rate, which is an input for the simulations (see Section 5.2). For all experiments, the pressurization phases can be simulated using the Flow-3D accommodation coefficients *rsize* summarized in Table 5.4. The column with the relative errors in Table 5.4 shows that for the majority of the GN2 pressurized simulations a maximal error of 8% can be achieved, which are very good results.

For the N200r and N400r experiments, which have the lowest pressurant gas temperature however, quite big errors, up to 48%, appear. This is due to the deviation of the numerical pressure curve from the experimental curve for these cases, as already depicted in Figure 5.4 (d). The reason for the deviation is assumed to be the calculation of the phase change mass flow rate.

The simulations of the GHe pressurized experiments show high errors compared to the experimental results. On one hand this is due to the fact that the pressurant gas masses are very small as it can be seen in the absolute errors. On the other hand is this due to the more complex numerical computation of the gas mixture with a non-condensable gas in the vapor phase. For all GHe simulations, the Flow-3D accommodation coefficient *rsize*, which is a multiplier on the phase change rate has to be set to 0.0001 to enable results (see Table 5.4). The simulations show a much faster increase in pressure than the experiments (see Figure 5.5), resulting in the high errors for the pressurant gas masses.

6.5 Amount of Mass Involved in Phase Change

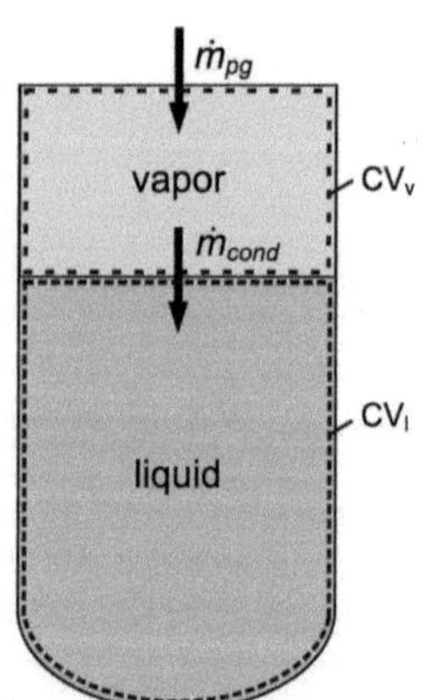

Fig. 6.10 Schematical propellant tank with vapor and liquid phase, pressurant gas and condensation mass flow rates as well as the applied control volumes.

Based on the results of the previous sections, analyses of the mass transport by phase change during the GN2 and GHe active-pressurization experiments will be presented hereafter. First the results of an analytical approach, based on experimental results, will be introduced. Numerical results will be presented afterwards.

6.5.1 Analytical Results

By means of the required pressurant gas mass, presented in Section 6.4, the phase change during the active-pressurization and the relaxation period can be assessed. It will later be shown that condensation predominates the pressurization and relaxation phases. On that account, the analytical approach used hereafter will be presented under the assumption of condensation as way of phase change. The amount of condensed GN2 at the end of pressurization and relaxation is analyzed for all performed experiments. Figure 6.10 depicts schematically the tank with the applied control volumes for the vapor phase (CV_v) and for the liquid phase (CV_l), marked by the dashed lines.

For the calculation of the vapor mass in the ullage, it is assumed that the temperature distribution between two consecutive temperature sensors in the experiments is linear. The validity of this assumption is based on the results of Section 6.3. By assuming saturation temperature at the liquid surface, the mean temperature in the ullage can be calculated by averaging over the internal energy, as presented in Section 2.2.5. It therefore follows from Equation 2.31 for the average vapor temperature $\bar{T}_v$

$$\bar{T}_v = \frac{c_v \pi R^2 \left(\int_{z_\Gamma}^{z_{T4}} T_v(z) \rho_v(z) dz + \int_{z_{T4}}^{z_{T5}} T_v(z) \rho_v(z) dz + \ldots + \int_{z_{T7}}^{z_{T13}} T_v(z) \rho_v(z) dz \right)}{c_v\, m_v} \tag{6.13}$$

where z_Γ is the height of the liquid surface, z_{T4} is the height of the temperature sensors T4 and so on. $T_v(z)$ is the linear temperature distribution between two temperature sensors and R is the inner radius of the tank. The density $\rho_v(z)$ can be calculated for each probe height from the corresponding temperature (according to the linear temperature distribution between two probes) and the pressure. Data of $\bar{T}_v$ for the experiments can be found in Table A.10 in the Appendix.

The vapor mass m_v can be calculated based on Equation 2.32.

$$m_v = \pi R^2 \left(\int_{z_\Gamma}^{z_{T4}} \rho_v(z) dz + \int_{z_{T4}}^{z_{T5}} \rho_v(z) dz + \ldots + \int_{z_{T7}}^{z_{T13}} \rho_v(z) dz \right) \tag{6.14}$$

For this calculation, the temperature sensor T14 next to the diffuser is disregarded, as it is highly dependent on the pressurant gas temperature.

For the analysis of the condensed mass, the vapor mass at the start of the pressurization (subscript "0"), at the end of pressurization (subscript "f") and after relaxation (subscript "T") are considered. Data of m_v for the experiments can be found in Table A.9 in the Appendix.

6.5.1.1 GN2 Pressurization

For the experiments pressurized with GN2, the vapor mass for all selected times can be calculated using Equation 6.14 (for the N300aH experiment with Tank Setup 2, it is modified accordingly with the additional temperature sensors). With the pressurant gas masses m_{pg}

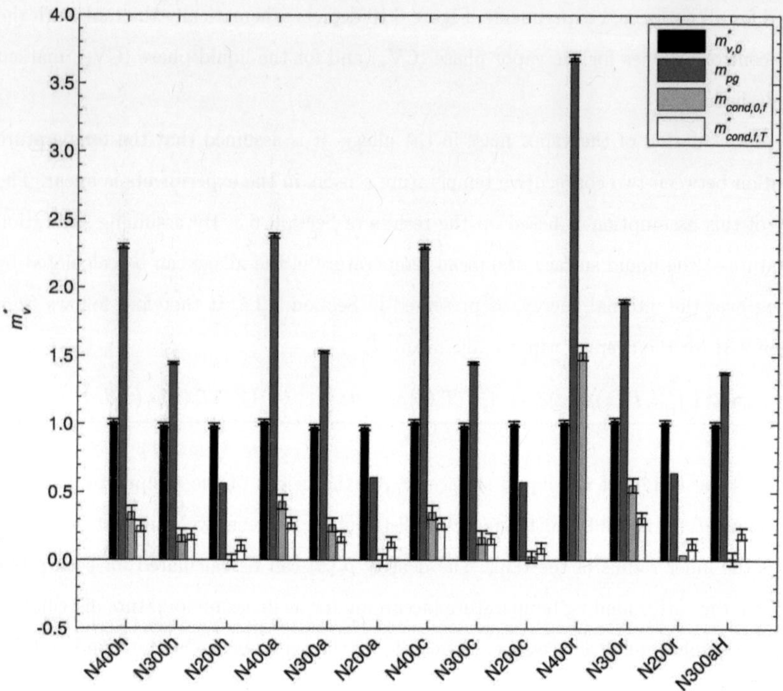

Fig. 6.11 Dimensionless vapor mass at pressurization start $(m_{v,0}^*)$, pressurant gas mass (m_{pg}^*), condensed vapor mass from pressurization start to pressurization end $(m_{cond,0,f}^*)$ and condensed vapor mass from pressurization end to relaxation end $(m_{cond,f,T}^*)$ for all GN2 pressurized experiments together with the relating errors. All data can be found in Table A.9 in the Appendix.

from Section 6.4, summarized in Table A.9 in the Appendix, the condensed masses of GN2 from the pressurization start until the end of the pressurization $m_{cond,0,f}$ can be calculated:

$$m_{cond,0,f} = m_{v,0} + m_{pg} - m_{v,f} \tag{6.15}$$

The condensed mass of GN2 from the pressurization end until the end of the relaxation $m_{cond,f,T}$:

$$m_{cond,f,T} = m_{v,f} - m_{v,T} \tag{6.16}$$

All results can be found in Table A.9 in the Appendix and are displayed in Figure 6.11.

In Figure 6.11, the amount of GN2 at pressurization start is depicted in dimensionless form by the black bars $(m_{v,0}^*)$. The nondimensional pressurant gas mass m_{pg}^* is depicted by the dark gray bars and supports the data, presented in the previous section: the longer the pressurization, the higher the required pressurant gas mass.

By regarding the amount of condensed GN2 from pressurization start until the pressurization end $m^*_{cond,0,f}$ (light gray bars), it can be seen, that the most GN2 condenses at the N400r experiment, which takes the longest to reach the final tank pressure (the pressurization durations t_{press} can be found in Table 6.2). For the N200h experiment, which has the shortest pressurization duration, the value for $m^*_{cond,0,f}$ is very small but negative, which means that this is the only experiment, for which evaporation dominates the phase change during pressurization. For the N200a experiment, $m^*_{cond,0,f}$ is nearly zero and for the N200c and N200r experiments, $m^*_{cond,0,f}$ has a small but positive value. It can also be seen that for the same pressurant gas temperatures, the amount of mass condensed during the pressurization phase increases with increasing final tank pressure level which correlates with the increased pressurization time. The comparison of $m^*_{cond,0,f}$ for the N300a and the N300aH experiments, which only differ in the fluid level, shows a decrease in pressurization time of about 10 s for the N300aH experiment, which has a higher liquid level. This results in a noticeably lower amount of condensed GN2 during the pressurization phase. This leads to the conclusion, that during very short pressurization phases, evaporation dominates, and the longer the pressurization phases, the more condensation takes place.

Table 6.2 Pressurization and relaxation times for the GN2 pressurized experiments.

$Exp.$	t_{press} [s]	t_{relax} [s]
N400h	96.8	254.1
N300h	60.7	170.3
N200h	23.6	64.1
N400a	100.1	251.9
N300a	64.3	148.3
N200a	25.5	55.2
N400c	96.8	241.0
N300c	60.9	178.9
N200c	24.1	55.5
N400r	155.2	-
N300r	79.9	331.3
N200r	27.0	41.9
N300aH	52.7	131.1

The total amount of condensed GN2 from the pressurization end until the end of relaxation, which is characterized by a horizontal pressure evolution (e.g. see Figure 6.1 (a)), is depicted by the white bars in Figure 6.11. Table 6.2 summarizes the relaxation times for the GN2 pressurized experiments, which are defined as $t_{relax} = t_{p,T} - t_{p,f}$. For the N400r experiment, there is no data available for t_{relax} and $m^*_{cond,f,T}$ as the experiment had to be aborted before the relaxation end.

In Figure 6.11, it can be seen that condensation is the dominant mode of phase change during the relaxation phase, even for the N200h experiment which has evaporation during the pressurization phase. The experiment with the highest amount of $m^*_{cond,f,T}$ is the N300r experiment, which also has the lowest pressurant gas temperature. Additionally it can be observed that the higher the final tank pressure, the more GN2 condenses afterwards. By comparing the results for the N300aH and the N300a experiments, it can be seen that although the N300aH experiment has less condensation in the pressurization phase, it has more condensation in the relaxation phase than the N300a experiment.

6.5.1.2 GHe Pressurization

For the helium pressurization, Equation 6.14 can only be used for the calculation of the vapor mass before pressurization. Afterwards, the ullage is filled with GN2 and GHe. On that account, the amount of GHe in the tank ullage after pressurization can be calculated with the molar mass of helium $M_{He} = 4.003{\cdot}10^{-3}$ kg/mol as follows.

$$n_{GHe} = \frac{m_{GHe}}{M_{He}} \tag{6.17}$$

For the vapor phase, an ideal gas mixture is assumed and that no helium is dissolved in the liquid phase.

In order to be able to analyze phase change of the helium pressurized experiments, the temperature of the free surface has to be determined for the times $t_{p,f}$ and $t_{p,T}$. At $t_{p,0}$ no helium is in the tank, so the saturation temperature can be calculated according to the tank pressure and the amount of GN2 in the tank ullage can be determined. However, during and after the GHe pressurization, the amount of evaporated or condensed nitrogen is not known from the start, and therefore the partial pressure of the GN2 cannot be specified. For his experiments, Arndt [3] had an additional temperature sensor floating on the free surface, measuring the temperature at this position. This solution was not applied for the experiments, presented in this study. The following approach was instead used to determine a saturation temperature for the times $t_{p,f}$ and $t_{p,T}$. In the first step, the mean ullage temperature $\bar{T}_v$ over the height of the vapor phase has to be estimated. At $t_{p,f}$, the lid temperature T13 is for this estimation disregarded, as the pressurant gas temperature measured at T14 is predominating the temperature profile. For $t_{p,T}$ the lid temperature T13 is used, but T14 is disregarded. The amount of substance in the tank ullage can therefore be calculated based on Equation 2.1.

$$n = \frac{p\,V_v}{\bar{R}\,\bar{T}_v} \tag{6.18}$$

Where n is the total amount of substance in the tank ullage, p is the tank pressure at the time $t_{p,f}$ or $t_{p,T}$, whichever is calculated, V_v is the ullage volume and $\bar{R}$ is the universal gas constant.

The amount of GN2 in the ullage can now be determined.

$$n_{GN2} = n - n_{GHe} \tag{6.19}$$

The mole fractions of GN2 and GHe can be calculated from Equation 2.3 as

$$\chi_{GN2} = \frac{n_{GN2}}{n} \tag{6.20}$$

and

$$\chi_{GHe} = \frac{n_{GHe}}{n}. \tag{6.21}$$

The partial pressure of helium and nitrogen in the tank ullage is calculated based on Equation 2.2.

$$p_{GN2} = p \, \chi_{GN2} \tag{6.22}$$

$$p_{GHe} = p \, \chi_{GHe} \tag{6.23}$$

The data for the GHe experiments is summarized in Tables A.4 and A.5 in the Appendix. With the partial pressure of nitrogen, the saturation temperature at that pressure can be calculated. Now the determination of the mean ullage temperature $\bar{T}_v$ over the height of the vapor phase has to be repeated with

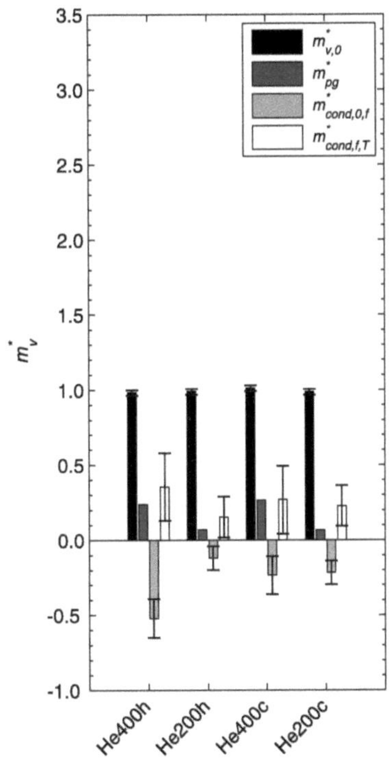

Fig. 6.12 Vapor mass at pressurization start $(m^*_{v,0})$, pressurant gas mass (m^*_{pg}), condensed vapor mass from pressurization start to pressurization end $(m^*_{cond,0,f})$ and condensed vapor mass from pressurization end to relaxation end $(m^*_{cond,f,T})$ for all GHe pressurized experiments together with the relating errors. All data can be found in Table A.9 in the Appendix.

the recently calculated saturation temperature. With this new ullage temperature, the calculations from Equation 6.18 to the determination of the new saturation temperature also have to be repeated. This iteration must be continued until the assumed mean ullage temperature corresponds to the calculated one. The next step checks whether the assumption of incompressibility in Equation 6.18 is correct. The results show, that for all helium experiments, the compressibility of nitrogen and helium corresponding to their partial pressures is equal or approximately 1. Consequently, the mass of GN2 in the vapor phase can be calculated as

$$m_{GN2} = n_{GN2} \, M_{N2} \tag{6.24}$$

with $M_{N2} = 28.013 \cdot 10^{-3}$ kg/mol as the molar mass of nitrogen. All results can be found in Table A.9 in the Appendix.

Table 6.3 Pressurization and relaxation times for the GHe pressurized experiments.

Exp.	t_{press} [s]	t_{relax} [s]
He400h	51.5	171.7
He200h	15.1	90.6
He400c	57.2	188.5
He200c	14.6	90.9

Figure 6.12 shows the relevant masses for all four helium pressurized experiments and Table 6.3 summarizes the pressurization and relaxation times. The GN2 mass at pressurization start $m_{v,0}$ is depicted with black bars and the GHe pressurant gas mass m_{pg} with dark gray bars. The condensed GN2 for the pressurization phase $m_{cond,0,f}$ has a negative value for all four experiments, which indicates that evaporation predominates during the GHe pressurization phase. One explanation for this is that the specific heat capacity c_p of gaseous helium is about 4.5 times as high as that of gaseous nitrogen (e.g. $c_{p,GHe} = 5.196{\cdot}10^3$ J/(kg K) and respectively $c_{p,GN2} = 1.124{\cdot}10^3$ J/(kg K) at $p = 101.3$ kPa and $T = 77.35$ K [62]). The injected helium has still a quite high temperature when it reaches the free surface and causes therefore evaporation. Based on the experimental results of the short GN2 pressurization phases, evaporation or hardly any phase change was determined. This might be due to the fact that the GN2 causes first evaporation when reaching the free surface and then gets cooled down faster than the GHe, resulting in condensation for the longer pressurization phases.

For the He400h experiment the most GN2 is evaporated, even though the required pressurant gas mass is very similar to that of the He400c experiment. The reason for that may be that in this experiment a short pressurant gas jet occurred at the beginning of the pressurization due to a mistake in operating. On that account, the pressure in the feed line increased and as V_2 was opened shortly after, the tank pressure increased for about three seconds very fast until the nominal pressure gradient appeared. That very quickly pressure increase might be the reason that a larger amount of GN2 evaporated for the He400h experiment until the pressurization end than would have been expected on the basis of the results of the other experiments. The condensed GN2 from pressurization end until relaxation end $m_{cond,f,T}$ is depicted for all four experiments with the white bars. It can be seen that, as for the GN2 pressurized experiments, condensation predominates during the relaxation phase. Please note that the data of Figure 6.12 and Table A.9 for $m_{cond,0,f}$ and $m_{cond,T}$ have an error of about 50 % due to inaccuracies in the estimation of the mean ullage temperature $\bar{T}_v$ at pressurization end and the relaxation end.

For the work of Arndt [3], the amount of helium dissolved in the LN2 was neglected. This assumption has to be verified for the current experiments. On that account, the concentration of the helium dissolved in the LN2 is calculated using the empirical correlation of van Dresar and Stochl [32] for concentration of helium dissolved in the liquid nitrogen:

$$\phi_{GHe} = 1.383 \cdot 10^{-10}(p - p_{sat})^{0.99} \, T_l^{3.82} \tag{6.25}$$

The pressure quantities have to be in MPa. The highest used tank pressure of the experiments is $p = 0.4$ MPa and the saturation temperature for the liquid temperature of $T_l = 77.6$ K is $p_{sat} = 0.104$ MPa. Therefore follows $\phi_{GHe} = 6.86 \cdot 10^{-4}$. For all four performed GHe experiments results a helium concentration ratio of $\phi_{GHe}/\chi_{GHe} \ll 1$, which means that, as for the work of Arndt [3], the helium dissolution in the LN2 can be neglected for this study.

6.5.2 Numerical Results

Table 6.4 Analytical and numerical phase change masses of the pressurization and relaxation phases. For the Flow-3D simulation of the relaxation phase of the N400r experiment (*), a relaxation time of 500 s is assumed. Errors are given of the Flow-3D results relative to the analytical masses. For other data see Table A.8 in the Appendix. Please note that all values are rounded.

	pressurization				relaxation			
	Analyt. [kg]	Flow-3D [kg]	error [%]	abs.error [kg]	Analyt. [kg]	Flow-3D [kg]	error [%]	abs.error [kg]
N200r	0.0014	0.0112	697.5	0.0098	0.0044	0.0109	147.2	0.0065
N300r	0.0194	0.0255	31.5	0.0061	0.0110	0.0063	-42.8	-0.0047
N400r	0.0534	0.0364	-31.9	-0.0170	-	-0.0291*	-	-
N200c	0.0009	0.0105	1062.5	0.0096	0.0032	0.0041	28.8	0.0009
N300c	0.0059	0.0220	272.4	0.0161	0.0056	0.0007	-88.2	-0.0049
N400c	0.0123	0.0238	93.8	0.0115	0.0095	-0.0057	-159.8	-0.0152
N200a	0.0000	0.0100	-50677.3	0.0100	0.0047	0.0089	89.6	0.0042
N300a	0.0091	0.0239	162.8	0.0148	0.0061	0.0050	-17.7	-0.0011
N400a	0.0150	0.0241	60.9	0.0091	0.0096	-0.0037	-138.9	-0.0133
N200h	-0.0001	0.0214	-21482.0	0.0215	0.0037	0.0051	38.9	0.0014
N300h	0.0064	0.0101	57.2	0.0037	0.0066	-0.0036	-153.8	-0.0102
N400h	0.0122	0.0239	95.9	0.0117	0.0087	-0.1510	-1835.7	-0.1597
N300aH	0.0004	0.0045	1027.4	0.0041	0.0071	-0.0104	-245.8	-0.0175
He200c	-0.0076	0.0028	-137.2	0.0104	0.0080	0.0282	252.1	0.0202
He400c	-0.0082	0.0154	-288.2	0.0236	0.0094	0.0607	545.4	0.0513
He200h	-0.0042	0.0010	-123.3	0.0052	0.0054	0.0306	466.7	0.0252
He400h	-0.0183	0.0067	-136.7	0.0250	0.0124	0.0398	221.1	0.0274

Table 6.4 summarizes the results of the Flow-3D simulations of the experiments for the phase change mass for the pressurization and the relaxation phases. The numerical results are determined from the Flow-3D output data for the phase change mass flux for each cell in

the height of the free surface, multiplied by the corresponding area of the free surface. The mean values of this phase change mass flow rates for each time are multiplied by the output time step and summed up for the considered pressurization or relaxation time resulting in the phase change masses, given in Table 6.4. The numerical results are compared to the analytical results, presented in the previous section. The pressurization times for the Flow-3D simulations are summarized in Table A.8 in the Appendix. For the relaxation times used for the results summarized in Table 6.4, the relaxation times of the experiments are used, which can also be found in Table A.8 in the Appendix. The N400r experiment had to be aborted before relaxation end and has therefore no data for this phase. For the Flow-3D simulation of the relaxation phase of the N400r experiment, a relaxation time of 500 s is assumed.

It can be seen in Table 6.4 that the simulation results are about the same order of magnitude as the analytical results. For the helium pressurized experiments however, the simulations do not confirm the analytical results, which state that during the pressurization phase evaporation is the dominating method of phase change. The numerical results for the phase change masses of the pressurization phase are assumed to be not very reliable due to the difficulties with the Flow-3D accommodation coefficient *rsize* and the parasitic currents (as already introduced in Chapter 5). In the Flow-3D calculations of relaxation phase the expected pressure drop occurs, the pressure calculated by Flow-3D however decreases much lower than that of the experiment and no horizontal pressure evolution appears at the analyzed time. More details on the numerical pressure drop will be given in Section 6.8.3. The analysis of the numerical phase change masses show however, that the results of Flow-3D for the active-pressurization and relaxation phases correspond to the analytical determinations in terms of magnitude but should not be used for further analyses.

6.6 Heat Transfer during Active-Pressurization and Relaxation

In the following section, the heat transfer during the active-pressurization and relaxation phases is analyzed. An energy balance is applied to the pressure and temperature data of the experiments and the results of numerical simulations for the heat transfer are presented.

6.6.1 Energy Balance

All data and values mentioned in this section can be found in Tables A.5, A.9 and A.10 in the Appendix. The energy balance for the tank system is based on the first law of thermodynamics for open systems, introduced in Section 2.2.5.2. If condensation predominates the phase change, as already presented in Section 6.5.1, and evaporation can be disregarded (as schematically depicted in Figure 6.13), Equation 2.36 takes the following form for the active-pressurization phase (index "$0, f$").

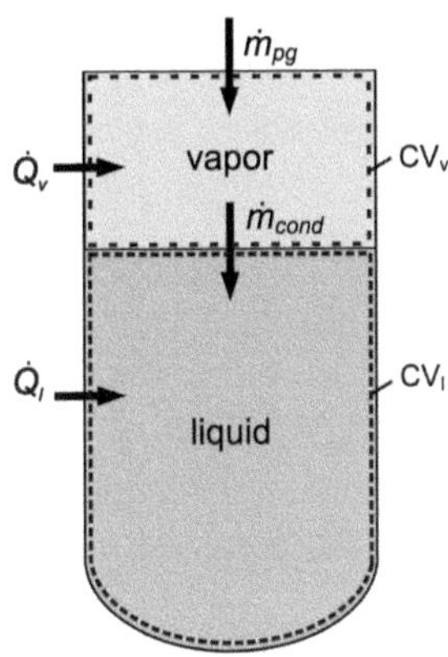

Fig. 6.13 Propellant tank with vapor and liquid phase, the reference control volumes and the considered heat and mass flow rates.

$$dQ_{l,0,f} = U_{l,f} - U_{l,0} - m_{cond,0,f} \, h_{v,cond,0,f} \qquad (6.26)$$

Where $m_{cond,0,f}$ is the condensed vapor mass from pressurization start until the pressurization end. It is calculated with Equation 6.15 presented in Section 6.5.1 (data in Table A.9 in the Appendix). In Equation 6.26, the mass entering the liquid control volume is assumed to be in liquid state. In order to convert that liquid mass into a vapor mass, the specific heat of vaporization is required. As the difference between the specific liquid and vapor enthalpy equals the specific heat of vaporization Δh_v it can be written: $h_{l,cond} + \Delta h_v = h_{v,cond}$. On that account, in Equation 6.26 the specific enthalpy of the vapor at surface level $h_{v,cond,0,f}$ is already used. $U_{l,f}$ is the internal energy of the total liquid at pressurization end and $U_{l,0}$ at the pressurization start. They can be written as

$$U_{l,0} = m_{l,0} \, u_{l,0} \qquad (6.27)$$

$$U_{l,f} = m_{l,f} \, u_{l,f} \qquad (6.28)$$

$$m_{l,f} = m_{l,0} + m_{cond,0,f} \qquad (6.29)$$

with the specific internal energy $u_{l,0}$ and the liquid mass $m_{l,0}$ at pressurization start, and $u_{l,f}$ and $m_{l,f}$ at pressurization end. The final mass of the liquid $m_{l,f}$ consists of the liquid mass at pressurization start $m_{l,0}$ and the condensed mass until the pressurization end $m_{cond,0,f}$.

Therefore, Equation 6.26 can be written as

$$dQ_{l,0,f} = m_{l,0}(u_{l,f} - u_{l,0}) + m_{cond,0,f}(u_{l,f} - h_{v,cond,0,f}) \qquad (6.30)$$

and accordingly for the relaxation phase (index "f, T")

$$dQ_{l,f,T} = U_{l,T} - U_{l,f} - m_{cond,f,T}\, h_{v,cond,f,T} \tag{6.31}$$

and

$$dQ_{l,f,T} = m_{l,f}(u_{l,T} - u_{l,f}) + m_{cond,f,T}(u_{l,T} - h_{v,cond,f,T}) \tag{6.32}$$

where $m_{cond,f,T}$ is the liquid mass condensed from pressurization end until relaxation end, calculated with Equation 6.16 (data in Table A.9 in the Appendix).

The liquid mass $m_{l,0}$ is calculated similarly to the calculation of the vapor mass with Equation 6.14. The temperature distribution between two sensors is assumed to be linear. For the simplification of the calculation, the bottom of the test tank is assumed to be flat at the height $z_b = 0.022$ m, which results in an identical tank volume as for the real test tank with the round shaped bottom. The mass $m_{l,f}$ is calculated as

$$m_{l,f} = m_{l,0} + m_{cond,0,f} \tag{6.33}$$

with the condensation vapor masses from Equation 6.15.

The specific internal energies u are determined using the NIST database [62] at the corresponding pressure and the average liquid temperature $\bar{T}_l$. The average temperature of the liquid phase is calculated with an integral over the liquid height

$$\bar{T}_l = \frac{\int_{z_b}^{z_{T8}} T_l(z)dz + \int_{z_{T8}}^{z_{T1}} T_l(z)dz + \ldots + \int_{z_{T3}}^{z_\Gamma} T_l(z)dz}{z_\Gamma - z_b} \tag{6.34}$$

where $T_l(z)$ is the linear temperature distribution between two sensors.

The enthalpies of the vapor at surface level $h_{v,cond,0,f}$ and $h_{v,cond,f,T}$ of Equations 6.30 and 6.32 are determined using the NIST database [62] with an average value of the tank pressure for the pressurization respectively the relaxation phase. The applied values are summarized in Tables A.9 and A.10 in the Appendix.

The energy balance for the vapor phase is based on Equation 2.37. By considering again only condensation for the pressurization and relaxation phase, it follows for the pressurization phase

$$dQ_{v,0,f} = U_{v,f} - U_{v,0} - m_{pg}\, h_{pg} + m_{cond,0,f}\, h_{v,cond,0,f} \tag{6.35}$$

resulting in:

$$dQ_{v,0,f} = m_{v,0}(u_{v,f} - u_{v,0}) + m_{cond,0,f}(h_{v,cond,0,f} - u_{v,f}) + m_{pg}(u_{v,f} - h_{pg}) \tag{6.36}$$

Accordingly for the relaxation phase, the energy balance can be written as

$$dQ_{v,f,T} = U_{v,T} - U_{v,f} + m_{cond,f,T}\ h_{v,cond,f,T} \tag{6.37}$$

and

$$dQ_{v,f,T} = m_{v,f}(u_{v,T} - u_{v,f}) + m_{cond,f,T}(h_{v,cond,f,T} - u_{v,T}). \tag{6.38}$$

The masses are calculated as already explained in Sections 6.4.1 and 6.5.1. The specific internal energies are, as for the liquid phase, determined using the NIST database [62] at the corresponding pressure and the average liquid temperature $\bar{T}_v$ (see Section 6.5.1). The specific enthalpy of the pressurant gas h_{pg} is determined by the average pressurant gas temperature, measured at T17 (corresponding to T_{pg}) and the average pressure, measured at p2 (see Figure 4.1, data summarized in Table A.8).

For the experiments with GHe as pressurant gas, the specific internal energy of a gas mixture is used. As example, the specific internal energy of the vapor phase at pressurization end is

$$u_{v,f} = \Psi_{GN2}\ u_{GN2} + \Psi_{GHe}\ u_{GHe} \tag{6.39}$$

where Ψ is the mass fraction, as introduced in Equation 2.4.

$$\Psi_{GHe,f} = \frac{m_{pg}}{m_{v,f}} \tag{6.40}$$

$$\Psi_{GN2,f} = \frac{m_{GN2,f}}{m_{v,f}} \tag{6.41}$$

All data are summarized in Table A.5 in the Appendix.

From the conservation of energy, the change in the heat of the fluids for the pressurization phase and the relaxation phase is determined. In Figure 6.14, the results of the energy balance from pressurization start until the pressurization end is depicted in four subplots for all GN2 pressurized experiments with Tank Setup 1. Each of the subplots represents one pressurant gas temperature and includes the results of the three experiments with that pressurant gas temperature. The different experiments are identified by their final tank pressure p_f. The black squares represent the total change in heat of the test tank over the pressurization phase $(dQ_{0,f})$. This value can be split up into the change in heat of the liquid phase $dQ_{l,0,f}$ and that of the vapor phase $dQ_{v,0,f}$. Additionally, the amount of energy of the pressurant gas $m_{pg}\ h_{pg}$ and the amount of energy, transferred from the vapor to the liquid phase by condensation $m_{cond,0,f}\ h_{v,cond,0,f}$ is shown.

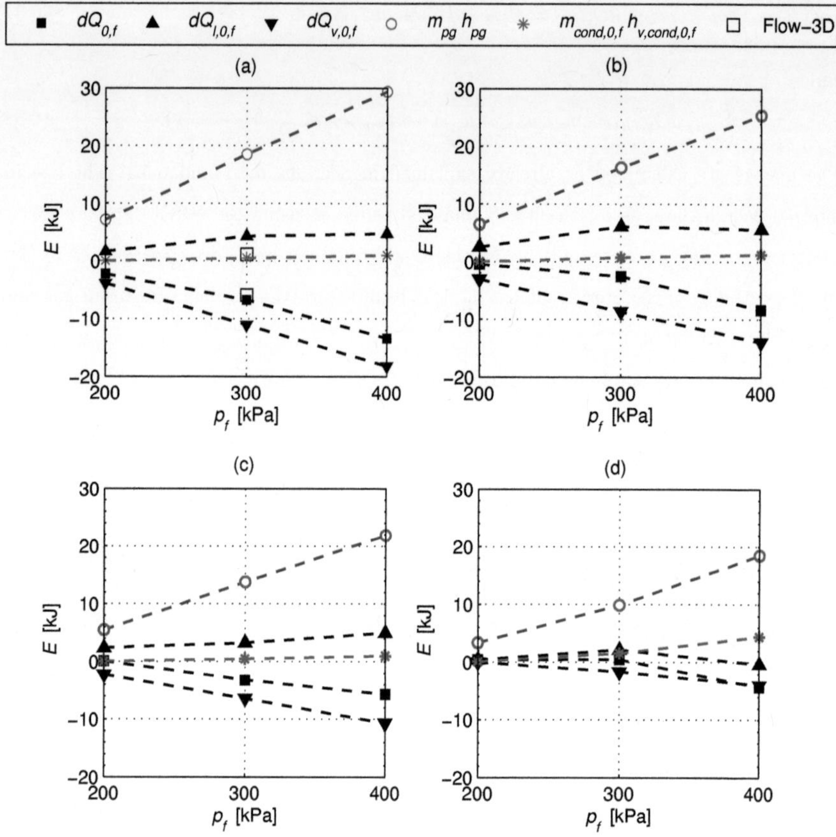

Fig. 6.14 Energy balance of the pressurization phase of GN2 pressurized experiments: (a) N200h, N300h, N400h, (b) N200a, N300a, N400a, (c) N200c, N300c, N400c, (d) N200r, N300r, N400r and Flow-3D results for the N300h experiment. The dashed lines are only for better visualization.

For all subplots of Figure 6.14, it can be seen that the change in heat for the total tank system $dQ_{0,f}$ has negative values. According to the definition of the signs in Section 6.6.1, this amount of heat leaves the tank system during the pressurization phase. By considering the results for the amount of energy transferred by condensation, it can be seen that the amount of energy that enters the liquid phase and leaves the vapor phase by phase change is very little. The only exception is the N400r experiment, which has, as already presented in Section 6.5.1.1, a relatively high amount of GN2 condensing over the pressurization phase. This leads to the conclusion that, during the pressurization period of the analyzed experiments, phase change has no major influence on the energy balance.

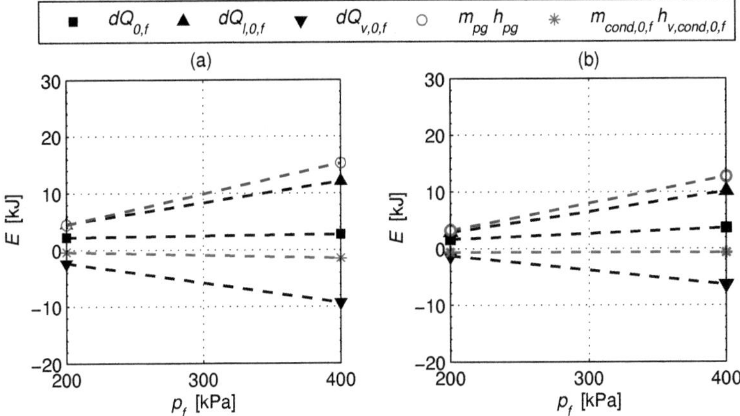

Fig. 6.15 Energy balance of the pressurization phase of GHe pressurized experiments: (a) He200h, He400h, (b) He200c, He400c. The dashed lines are only for better visualization.

Comparatively, the energy of the pressurant gas $m_{pg}h_{pg}$ has a much bigger impact then the phase change on the energy balance It primarily affects the vapor phase. The high amount of energy of the pressurant gas leads, according to Equation 6.36, to high, but negative values for the change in heat of the vapor phase $dQ_{v,0,f}$. This means, that the amount of energy brought into the system by the pressurant gas leaves the vapor phase very fast. As the amount of energy, transferred over the free surface by condensation has hardly any impact, it is assumed that the majority of the heat leaving the vapor phase, goes into the tank wall and warms it up. By comparing the different pressurant gas temperatures, it can be seen that, for the highest pressurant gas temperature (Figure 6.14 (a)) the most heat leaves the vapor phase and for the lowest pressurant gas temperature (Figure 6.14 (d)) the least heat leaves for the performed experiments.

Based on this it can be stated that the change in heat of the liquid phase $dQ_{l,0,f}$, is dominated by the heat entering from the wall. This explains the positive sign of $dQ_{l,0,f}$ in Figure 6.14. Moreover, if we recall the evolution of the tank wall temperatures of Figures 6.1 (d) and 6.2 (d), it can be seen that this assumption is valid: in Figure 6.1 (d), the sensors T9 to T12 are situated only in the vapor phase, whereas in Figure 6.2 (d) sensor T9 is in the liquid phase at 0.005 m below the free surface. Sensors T9 to T12 all show increasing wall temperatures over the pressurization period. As the lid temperature (T13) stays constant over time, all heat leaving the vapor phase enters the vertical tank walls. A small part of it is conducted down to the liquid phase, heating up the uppermost liquid layers.

Figure 6.15 shows the energy balance for the pressurization phase of the helium pressurized experiments. The main difference between the change in heat for the GN2 and the GHe pressurization is the phase change: during the GN2 pressurization, condensation is the dominating way of phase change and for the helium pressurization it is evaporation (as already presented in Section 6.5.1.2). In order to maintain the definition of the signs, introduced in Section 6.6.1, condensation is defined as the positive way of phase change. On that account, the energy transferred from the vapor to the liquid phase by condensation $m_{cond,0,f}\,h_{v,cond,0,f}$ has negative values. But, as already for the GN2 pressurization, also in this case the impact of the phase change on the energy balance is of minor importance. The energy of the pressurant gas $m_{pg}h_{pg}$ has the highest impact on the energy balance of the pressurization phase, dependent on the pressurant gas temperature.

By comparing Figure 6.15 with Figures 6.14 (a) and (c), it can be seen that the energy of the pressurant gas $m_{pg}h_{pg}$ for the GHe pressurized experiments is noticeably lower than that of the GN2 pressurized experiments. This is due to the fact that the pressurant gas mass of the GHe experiments is much lower than that of the GN2 experiments, even though the specific enthalpy is considerably higher for helium (see Table A.9). Moreover, the change in heat of the liquid phase is higher for the GHe experiments, as evaporation is the dominating mode of phase change.

Figure 6.16 depicts the energy balance of the relaxation phase for the GN2 pressurized experiments with Tank Setup 1. As already mentioned, there is no data for the N400r experiment for the relaxation phase. For the relaxation period the amount of energy, transferred from the vapor to the liquid phase by condensation $m_{cond,f,T}\,h_{v,cond,f,T}$ has only little impact on the energy balance of the relaxation period. The vapor phase shows only few change in heat, compared to the pressurization phase, as only a small amount of heat leaves the vapor phase and enters the tank wall. Figure 6.2 (d) shows the evolution of the wall temperature during the relaxation phase and it can be seen that the wall temperature in the liquid phase increases over time whereas the wall temperature in the vapor phase decreases. This observation correlates with the results of the change in heat of the liquid phase $dQ_{l,f,T}$, as there was a larger amount of heat entering the liquid phase during relaxation. The heat, which enters the wall through the vapor phase during pressurization is conducted downwards to

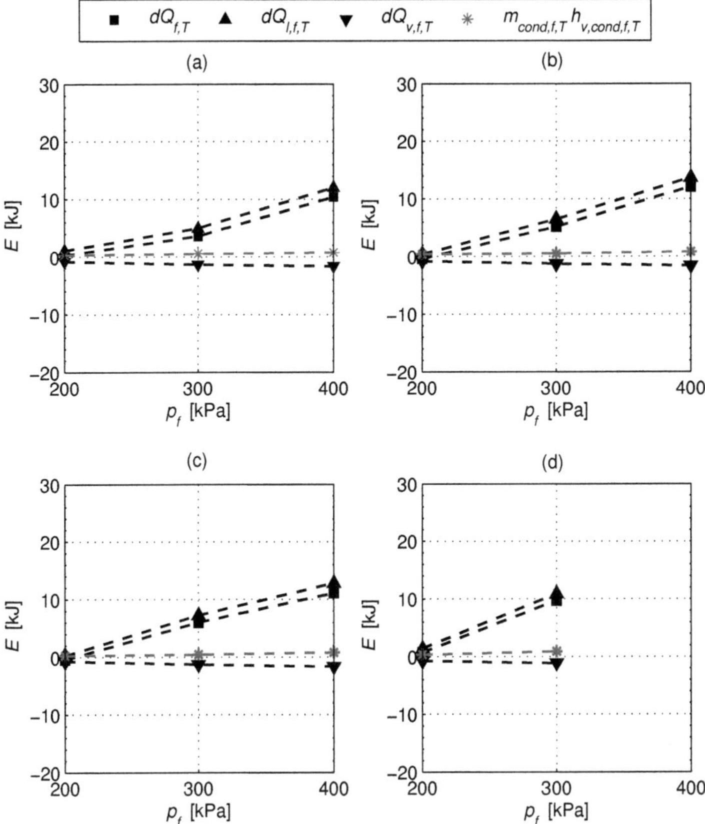

Fig. 6.16 Energy balance of the relaxation phase of GN2 pressurized experiments: (a) N200h, N300h, N400h, (b) N200a, N300a, N400a, (c) N200c, N300c, N400c, (d) N200r, N300r. The dashed lines are only for better visualization.

the liquid. As the pressurization phases are very short, the heat conduction still continues during the relaxation phase. It is assumed that the heat enters the liquid phase right below the free surface, warming up the uppermost liquid layers. Now heat conduction takes also place in the liquid phase in a downward direction. Comparing all four subplots of Figure 6.16 shows, that during the relaxation phase only little difference appears in the energy balance between the results for the different pressurant gas temperatures.

Figure 6.17 displays the energy balance of the GHe pressurized experiments for the relaxation period. In contrast to the pressurization phase it can be seen that during the relaxation

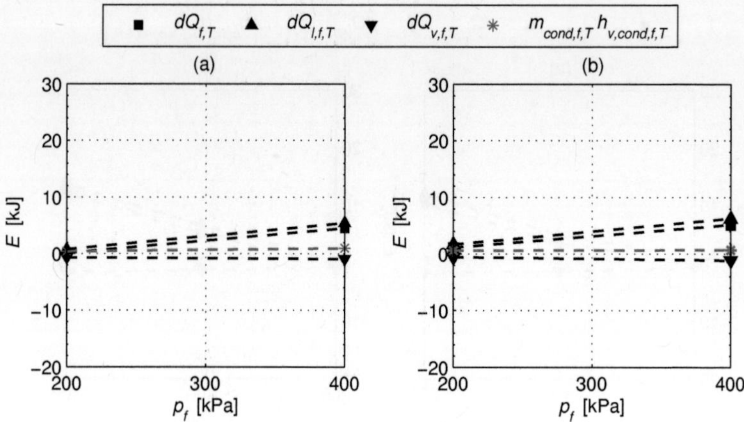

Fig. 6.17 Energy balance of the relaxation phase of GHe pressurized experiments: (a) He200h, He400h, (b) He200c, He400c. The dashed lines are only for better visualization.

phase of the helium pressurized experiments, condensation is the predominating method of phase change (see Section 6.5.1.2). The energy, transferred over the free surface by condensation $m_{cond,f,T}\, h_{v,cond,f,T}$ has therefore positive values, which are again very small. The change in heat of the vapor phase also indicates only a small heat transfer to the tank wall. As during the pressurization phase, less energy of the pressurant gas $m_{pg} h_{pg}$ is brought into the vapor phase compared to the GN2 experiments and hence the change in heat for the liquid phase has smaller values compared to Figure 6.16 (a) and (c). Nevertheless, during the relaxation period the same effects of heat conduction through the wall appear for the GHe pressurized experiments as for the GN2 experiments.

6.6.2 Numerical Results

The heat transfer of the pressurization phase can also be analyzed using the results of the Flow-3D calulations. In Figure 6.14 (a), the Flow-3D results for the change in heat of the vapor and the liquid phase of the N300h experiment are depicted with the white squares. This change in heat is calculated separately for the vapor and the liquid phase with the Flow-3D results for the wall to fluid heat flux (Figure 6.19), multiplied by the corresponding wall area and the pressurization time. In Figure 6.14 (a), the upper white square corresponds to the change in heat of the liquid phase $dQ_{l,0,f} = 1.167 \cdot 10^3$ J and the lower white square to the change in heat of the vapor phase $dQ_{v,0,f} = -5.942 \cdot 10^3$ J. By comparison of the

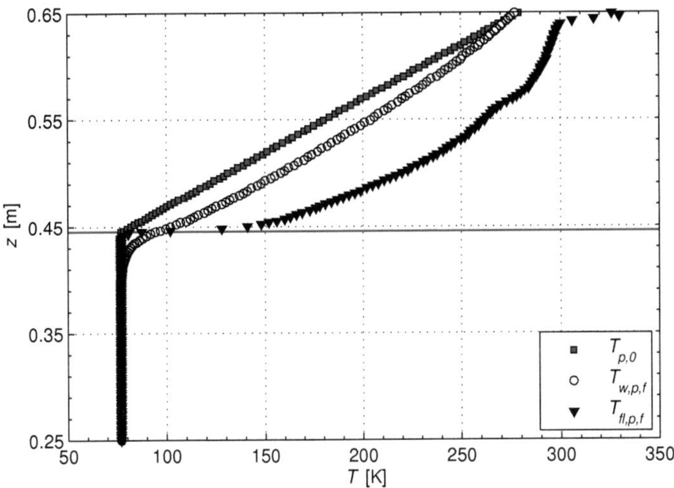

Fig. 6.18 Flow-3D results for the N300h experiment: wall and fluid temperature at pressurization start $(T_{p,0})$ and wall temperature $T_{w,p,f}$ and fluid temperature $T_{fl,p,f}$ at pressurization end over the tank height z. The free surface is at $z = 0.445$ m and the data for the fluid temperature $T_{fl,p,f}$ are taken at $r = 0.035$ m.

numerical results and the results of the energy balance it can be seen that the change in heat, calculated by Flow-3D has smaller values. This might be due to the fact that only heat transfer over the tank wall is considered for this analysis. Moreover, also the difference in the phase change mass, presented in Section 6.5.2, and the corresponding heat transfer might contribute to the error.

In the following, the numerical results of the N300h experiment are discussed. In the previous section, it was stated that the heat transfer over the tank wall provides the biggest contribution to heat transfer during the active-pressurization phase. Figure 6.18 depicts the temperature distribution of the Flow-3D simulation over the tank height z. The free surface is marked with the horizontal line at $z = 0.445$ m. Only the upper part of the liquid phase is depicted for improved representation, as the data is not changing below $z = 0.25$ m. The experimental temperature data at pressurization start is imported as initial temperature distribution for the simulations. The temperature 280 K ($z = 0.65$ m). In Figure 6.18, is set equal for the fluids and the tank wall ($T_{p,0}$ in Figure 6.18). After pressurization start, the wall is affected by heat transfer from the fluids and the lid, which has a constant temperature

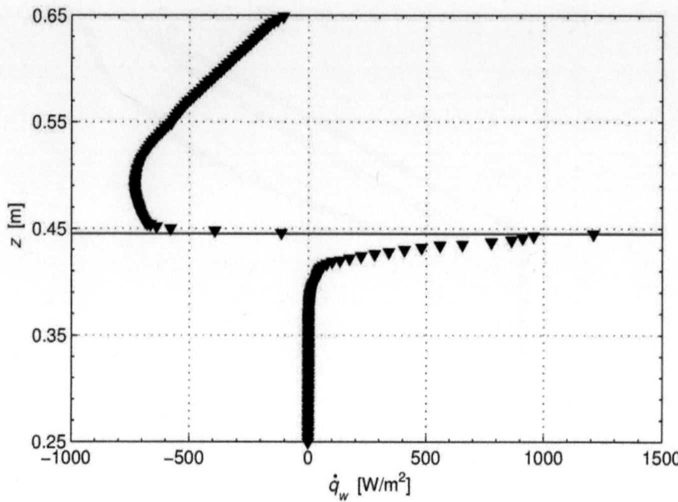

Fig. 6.19 N300h experiment: Flow-3D results of the wall to fluid heat flux $\dot{q}_w$ at pressurization end over the tank height z. The free surface is at $z = 0.445$ m.

of the temperature distribution at pressurization end is shown for the tank wall ($T_{w,p,f}$) and the vapor, respectively the liquid phase $T_{fl,p,f}$ at $r = 0.035$ m. In the upper part of the vapor temperature, the flow of the hot pressurant gas is visible. It can be seen that the wall gets heated up in the region of the vapor phase ($z = 0.445$ m to $z = 0.65$ m) and also in the upper region of the liquid phase. This is also confirmed by the experimental data: in Figure 6.2 (d), the wall temperature at sensor T9, which is placed in the liquid phase for this experiment, increases during the pressurization phase but with a different gradient than the sensors T10 to T12, placed in the vapor phase at the wall.

Figure 6.19 depicts the Flow-3D results of the wall to fluid heat flux of the N300h experiment at pressurization end. The position of the tank wall is at $\dot{q} = 0$ W/m^2. It can be seen that in the region of the vapor phase the wall to fluid heat flux has negative values, which indicates that heat leaves the vapor phase and goes into the tank wall. The absolute value of the heat flux in the vapor area increases from the lid toward the free surface. The maximal absolute value is at the position where the maximal temperature difference appears between the vapor and wall temperature, as depicted in Figure 6.18. Below that point, the absolute values of the wall to fluid heat flux decreases again. At the position of the free surface ($z = 0.445$ m) the maximal value of the wall to fluid heat flux appears. Underneath

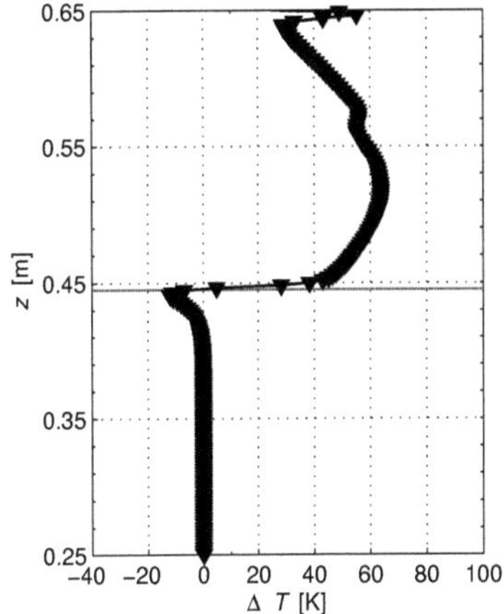 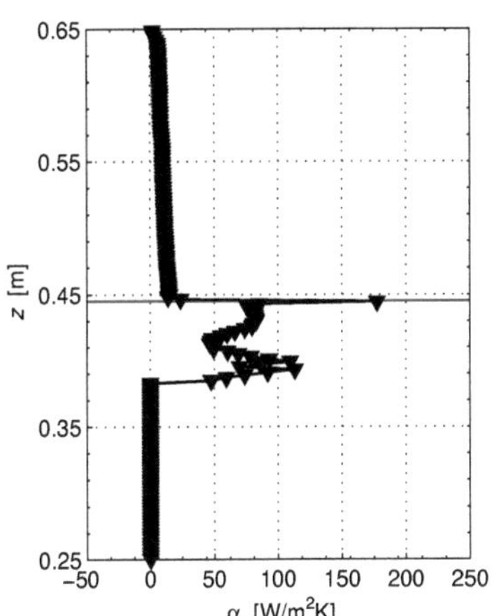

Fig. 6.20 N300h experiment: Temperature difference between the tank wall temperature and the fluid temperature at pressurization end over the tank height z ($\Delta T = T_{fl,p,f} - T_{w,p,f}$). The lines between the data points are only for better visualization and the free surface is at $z = 0.445$ m.

Fig. 6.21 N300h experiment: Heat transfer coefficient α based on the Flow-3D results at pressurization end over the tank height z. The lines between the data points are only for better visualization and the free surface is at $z = 0.445$ m.

the free surface, heat enters the liquid phase from the tank wall with decreasing amount in lateral direction, which is again due to the decreasing temperature difference between the liquid propellant and the tank wall. The thermal boundary layer only exists in the topmost liquid layers (see Section 6.3). Figure 6.19 underlines the assumption that the tank wall in the region of the liquid phase is heated up during the pressurization phase due to heat entering the wall at the height of the vapor phase which is then conducted downwards, heating up the liquid next to the tank wall right below the free surface.

From the results of Flow-3D for the wall to fluid heat flux and the temperature difference between the fluid and the wall temperature, depicted in Figure 6.20, the heat transfer coefficient of the tank wall can be determined analytically. Figure 6.21 depicts the heat transfer coefficient α of the tank wall, as introduced in Formula 2.45. It is determined with the data from Figures 6.19 and 6.20. The heat transfer coefficient has small values over the total ullage height, which increase slightly toward the free surface. In order to determine the mode of convection appearing in the presented case, an empirical correlation for the average NUS-SELT number for forced convection is applied. Therefore the following fluid parameters are used for the pressure of 300 kPa and a mean vapor temperature of 181.3 K from the NIST database [62]: $\nu_v = 2.12 \cdot 10^{-6}$ m^2/s, $D_{t,v} = 2.83 \cdot 10^{-6}$ m^2/s, resulting in a PRANDTL number of $\mathbf{Pr_v} = \nu_v/D_{t,v} = 0.75$. With the mean vertical flow velocity from the Flow-3D calulations $\bar{v}_z = 0.123$ m/s and the vertical tank wall height as characteristic length $L_v = 0.205$ m follows for the REYNOLDS number $\mathbf{Re_v} = (\bar{v}_z L_v)/\nu_v = 11894$. According to Cengel [19] applies for laminar forced convection flows Equation 2.49: $\mathbf{Nu_v} = 0.664\,\mathbf{Re_v}^{1/2}\mathbf{Pr_v}^{1/3}$ for $\mathbf{Re} < 10^5$. With $\lambda_v = 0.017$ W/(m K) results from the NUSSELT number $\mathbf{Nu_v} = (\alpha_v L_v)/\lambda_v$ a heat transfer coefficient of $\alpha_v = 5.4$ W/(m^2 K). The mean value from the numerical calulations depicted in Figure 6.21 is $\alpha_v = 10.2$ W/(m^2 K). Based on these numerical results can be stated that the flow of the vapor phase of the presented test case during pressurization can be assumed as a laminar forced convection flow.

In the liquid phase, a heat transfer coefficient appears only for a certain fluid layer thickness below the free surface. This confirms that a thermal stratified layer in the upper part of the liquid phase appears. In the Flow-3D calculation presented here, the thickness of that boundary layer in Figure 6.21 is 0.06 m. To identify the mode of convection, which appears, an empirical correlation for the average NUSSELT number for natural convection over surfaces is applied. According to Equation 2.48, the tank can be considered as a vertical plate. For the NUSSELT correlation the following data are used at 300 kPa and a mean liquid temperature in this thermal boundary layer of $\bar{T}_l = 79.1$ K from the NIST database [62]: $\nu_l = 1.89 \cdot 10^{-7}$ m^2/s, $D_{t,l} = 8.65 \cdot 10^{-8}$ m^2/s, resulting in a PRANDTL number of $\mathbf{Pr} = \nu_l/D_{t,l} = 2.18$. With $\beta_{T,l} = 5.66 \cdot 10^{-3}$ 1/K at the mean fluid temperature of 77.5 K, the characteristic lenght $L_l = 0.06$ m and $\Delta T_l = 3.06$ K, which is the mean temperature difference over the considered thermal boundary layer, a GRASHOF number of $\mathbf{Gr} = (g\,\beta_{T,l}\,L^3\,\Delta T_l)/\nu_l^2 = 10.32 \cdot 10^8$ results. It therefore follows a RAYLEIGH number of $\mathbf{Ra} = \mathbf{Pr}\,\mathbf{Gr} = 22.49 \cdot 10^8$. According to Cengel [19] applies for laminar free convection flows Equation 2.46: $\mathbf{Nu} = 0.59\,\mathbf{Ra}^{1/4}$ for $10^4 < \mathbf{Ra} < 10^9$. With $\lambda_l = 0.142$ W/(m K) results from the NUSSELT number $\mathbf{Nu} = (\alpha_l\,L_l)/\lambda_l$ a mean heat transfer coefficient of $\alpha_l = 304.1$ W/(m^2 K). The mean value from the numerical calulations depicted in Figure 6.21 is $\alpha_l = 73.2$ W/(m^2 K), neglecting the part where $\alpha_l \approx 0$ W/(m^2 K). Calculating the wall to fluid heat flux, using the heat transfer coefficient from the NUSSELT correlation and the mean temperature difference, results in $\dot{q}_w = \alpha_l\,\Delta T_l = 930.5$ W/m^2. By comparison with Figure 6.19, it can be seen that this result is valid for the uppermost 0.01 m of the liquid. It is assumed that the quite big difference between the analytical and the numerical results for

α_l is due to the fact that the heat, which is conducted in the tank wall further downwards is disregarded in the analytical approach, but covered by the Flow-3D simulations. However, based on these numerical results can be supposed that for the presented test case laminar free convection flow dominates near the tank wall in the top layers of the liquid phase during the pressurization period.

6.7 Correlation for the Pressure Rise

Based on the results of the previous sections, a correlation is presented hereafter, which can be used as an a priori approximation of the pressure change during the pressurization phase. The correlation is based on the ideal gas law (Equation 2.1) and for pressurization with helium, an ideal gas mixture is supposed. Furthermore, an isochoric system is assumed and phase change is disregarded (see Section 6.6.1). It therefore follows for the change in tank pressure:

$$\frac{dp(t)}{dt} = \frac{R_{s,0}\bar{T}_{v,0}}{V_u}\frac{dm_v(t)}{dt} + \frac{p_0}{\bar{T}_{v,0}}\frac{d\bar{T}_v(t)}{dt} + \frac{p_0}{R_{s,0}}\frac{dR_s(t)}{dt} \tag{6.42}$$

Therefore, the ullage volume V_u and the initial tank pressure p_0 have to be known. The change in vapor mass is only due to the pressurant gas mass flow rate $\dot{m}_{pg}$, which also has to be known a priori.

The change in the mean temperature of the vapor phase $\bar{T}_v(t)$ can be determined with the first law of thermodynamics for open systems applied for the vapor phase over the pressurization period (Equation 6.37), disregarding phase change, and using the definition of the internal energy of an ideal gas (Equation 2.29).

$$m_{v,f}c_{v,f}\bar{T}_{v,f} = dQ_{v,0,f} + m_{v,0}c_{v,0}\bar{T}_{v,0} + m_{pg}h_{pg} \tag{6.43}$$

$$\bar{T}_{v,f} = \frac{m_{v,0}c_{v,0}\bar{T}_{v,0} + m_{pg}h_{pg} + dQ_{v,0,f}}{m_{v,f}\,c_{v,f}} \tag{6.44}$$

Therefore, the initial vapor mass $m_{v,0}$, the average initial vapor temperature $\bar{T}_{v,0}$ and the specific enthalpy of the pressurant gas h_{pg} have to be known. For the nitrogen pressurized experiments, the specific heat capacity c_v can be assumed as constant over the pressurization phase $c_{v,0} = c_{v,f} = c_{v,GN2} = 745$ J/(kg K). For the helium pressurized cases $c_{v,0} = c_{v,GN2}$, during pressurization however, $c_{v,f}$ has to be determined by means of the mass fractions (Equation 2.4) in the vapor phase with $c_{v,GHe} = 3116$ J/(kg K).

$$c_v = \sum_i \Psi_i c_{v,i} \tag{6.45}$$

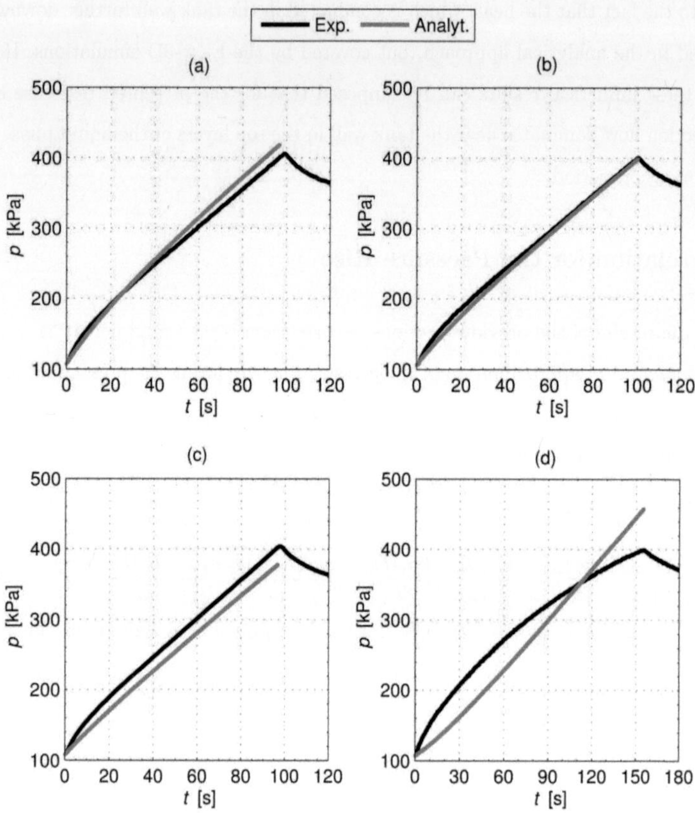

Fig. 6.22 Comparison of experimental pressure curve and results of the correlation of Equation 6.49 for the (a) N400h (b) N400a (c) N400c and (d) N400r experiments.

In Equation 6.44, the change in heat in the vapor phase $dQ_{v,0,f}$ between pressurization start and end is dominated by the heat transfer by convection between the vapor phase to the tank wall (as presented in Section 6.6.2) and can therefore be determined, based on Equation 2.45, as

$$\dot{Q}_v = \alpha A_w \Delta T_{v,w} \tag{6.46}$$

with the wall area in the tank ullage $A_w = 0.19$ m^2 ($H_v = 0.205$ m). The heat transfer coefficient α can be determined by means of the NUSSELT correlation, as already presented in Section 6.6.2. For the calculation of the required parameters, the mean final vapor temperature and the final tank pressure are needed. As final tank pressure for the helium pressurized experiments, the partial pressure of helium at pressurization end has to be presumed. For

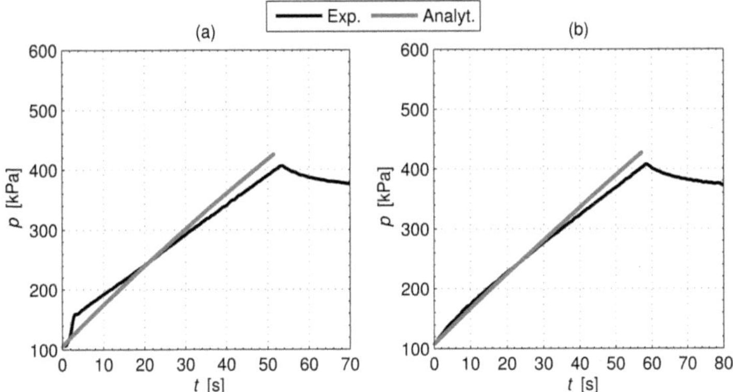

Fig. 6.23 Comparison of experimental pressure curve and results of the correlation of Equation 6.50 for the (a) He400h and (b) He400c experiments.

the analysis presented hereafter $p_{f,GHe} = 200$ kPa is assumed. The mean vapor temperature at pressurization end is determined as the mean value between the mean pressurant gas temperature at the diffuser $\bar{T}_{dif}$, and the saturation temperature at pressurization end $T_{sat,f}$. For the performed experiments $\bar{T}_{dif}$ was determined by means of experimental data as: N400h experiment $\bar{T}_{dif} = 315.1$ K, N400a experiment $\bar{T}_{dif} = 283.3$ K, N400c experiment $\bar{T}_{dif} = 270.0$ K and N400r experiment $\bar{T}_{dif} = 211.9$ K. This temperature is the mean temperature of the injected pressurant gas at temperature sensor T14 (compare Figures 4.2 and 6.7) and is selected as it represents the maximal possible vapor temperature, whereas $T_{sat,f}$ is the minimal possible vapor temperature at pressurization end (for $p_f = 400$ kPa $T_{sat,f} = 91.2$ K).

With the mean vertical flow velocity from the Flow-3D calculations $\bar{v}_z = 0.123$ m/s and the tank wall height in the ullage as characteristic length $L_v = 0.205$ m, a mean value of $\alpha = 6.3$ W/(m^2 K) for the GN2 pressurized experiments and of $\alpha = 9.5$ W/(m^2 K) for the GHe pressurized experiments is determined, which is taken as constant over the pressurization phase.

For Equation 6.46, the temperature difference $\Delta T_{v,w}$ has to be determined. Therefore, the maximal possible temperature difference between vapor phase and tank wall is used.

$$\Delta T_{v,w} = \bar{T}_w - \bar{T}_{dif} \tag{6.47}$$

With the mean wall temperature $\bar{T}_w$, determined as the mean value between $\bar{T}_{dif}$ and $T_{sat,0}$, the saturation temperature at pressurization start which corresponds to the minimal possible wall temperature. For the mean initial pressure of all experiments $p_0 = 106$ kPa follows $T_{sat,0} = 77.7$ K [62].

The change in the specific gas constant of Equation 6.42 needs to be considered only for the helium pressurized cases. It is the difference between the specific gas constant of the gas mixture at pressurization end and the specific gas constant of GN2 at pressurization start $R_{s,GN2} = 296.8$ J/(kg/K). The specific gas constant of an ideal gas mixture is defined by means of the mass fractions (Equation 2.4) as

$$R_s = \sum_i \Psi_i R_{s,i} \tag{6.48}$$

with $R_{s,GHe} = 2077$ J/(kg/K).

Implementing all this into Equation 6.42 and with integration over the pressurization phase from $t_{p,0}$ to t results for the GN2 pressurized experiments in the following correlation for the pressure rise.

$$p(t) = \frac{R_{s,GN2}\bar{T}_{v,0}}{V_u} \; \dot{m}_{pg}h_{pg}(t - t_{p,0}) + \frac{p_0}{\bar{T}_{v,0}} \tag{6.49}$$

$$\left\{ \frac{m_{v,0}c_{v,GN2}\bar{T}_{v,0} + \dot{m}_{pg}(t - t_{p,0}) + \left[\alpha A_w \left(\left(\frac{\bar{T}_{dif} - T_{sat,0}}{2} + T_{sat,0} \right) - \bar{T}_{dif} \right) \right](t - t_{p,0})}{(m_{v,0} + \dot{m}_{pg}(t - t_{p,0}))\, c_{v,GN2}} \right\}$$

And for the helium pressurized experiments follows for the correlation for the pressure rise:

$$p(t) = \frac{R_{s,GN2}\bar{T}_{v,0}}{V_u} \; \dot{m}_{pg}h_{pg}(t - t_{p,0}) + \frac{p_0}{\bar{T}_{v,0}} \tag{6.50}$$

$$\left\{ \frac{m_{v,0}c_{v,GN2}\bar{T}_{v,0} + \dot{m}_{pg}(t - t_{p,0}) + \left[\alpha A_w \left(\left(\frac{\bar{T}_{dif} - T_{sat,0}}{2} + T_{sat,0} \right) - \bar{T}_{dif} \right) \right](t - t_{p,0})}{(m_{v,0} + \dot{m}_{pg}(t - t_{p,0})) \left[\frac{m_{v,0}}{m_{v,0} + \dot{m}_{pg}(t - t_{p,0})}c_{v,GN2} + \left(1 - \frac{m_{v,0}}{m_{v,0} + \dot{m}_{pg}(t - t_{p,0})} \right) c_{v,GHe} \right]} \right\}$$

$$+ \frac{p_0}{R_{s,GN2}} \left[\frac{m_{v,0}}{m_{v,0} + \dot{m}_{pg}(t - t_{p,0})}R_{s,GN2} + \left(1 - \frac{m_{v,0}}{m_{v,0} + \dot{m}_{pg}(t - t_{p,0})} \right) R_{s,GHe} - R_{s,GN2} \right]$$

Figures 6.22 and 6.23 show the results of Equations 6.49 and 6.50 compared to the experimental pressure rises. It can be seen that for the N400h, N400a and N400c test cases, the results of the correlation agree very well with the experimental data, with a maximal error of 6 %, confirming the assumptions made. For the N400r experiment, which has the lowest pressurant gas temperature of 144 K, the experimental pressure curve is not covered very well. It is assumed that for such low pressurant gas temperatures, phase change should not be disregarded, as it represents a considerable amount of change in energy (see Figures 6.11 and 6.14 (d)). For the helium pressurized test cases, a maximal error of 7 % between the results of the correlation and the experiments appears, which is also a very good result and verifies the assumptions made for the gas mixture. For GHe pressurized test cases, the partial pressure of helium has to be presumed, which makes it challenging to obtain very accurate results. For the results in Figure 6.23, $p_{f,GHe} = 200$ kPa is used. By using the partial pressures, summarized in Table A.4 in the Appendix, which were determined a posteriori based on the experimental results, a maximal error of 5 % between the correlation and the experimental results can be obtained. Please note that the presented correlations are only validated for the presented experiments and test setup.

6.8 Pressure Drop after Pressurization End

As the tank pressure rise by the active-pressurization is finished, the experiments show a subsequent drop in the pressure which is analyzed in this section: first, the experimental results are presented, then results of a theoretical approach are introduced and afterwards numerical results are given.

6.8.1 Experimental Pressure Drop

In the active-pressurization experiments, the test tank is pressurized up to the final tank pressure. After this is reached, the pressurant gas inflow is stopped and the tank is let closed until the experiment is completed. As the pressurant gas inflow is stopped, the tank pressure decreases instantly and shows an asymptotical evolution until that point which is defined as relaxation end. If the tank is still kept closed after that point, the tank pressure increases again. In the performed experiments, three final tank pressures and two different pressurant gases were applied and for all experiments, the pressure drop appeared.

Figure 6.24 shows the evolution of the tank pressure during the N300h active-pressurization experiment (details see Table A.8 in the Appendix). Section (A) is the pressurization phase, section (B) the relaxation phase and section (C) is the pressure increase after relaxation

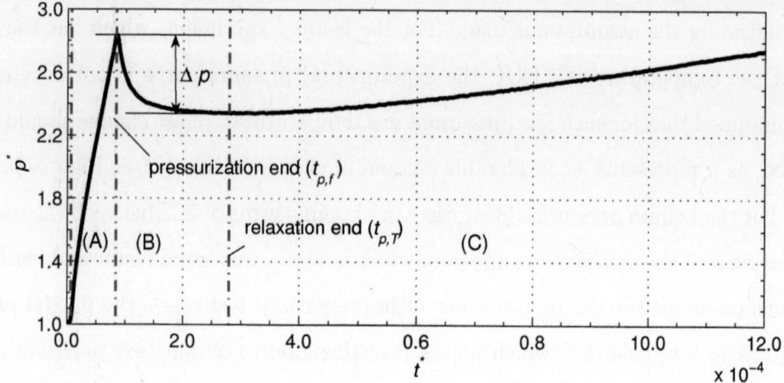

Fig. 6.24 Evolution of the tank pressure of the N300h experiment: (A) pressurization, (B) relaxation, (C) pressure increase after relaxation (Tank Setup 1, see Figure 4.2, detailed data in Table A.8 in the Appendix). Pressurization starts at $t_{p,0}$ ($t^* = 0.06 \cdot 10^{-4}$) and ends at $t_{p,f}$ ($t^* = 0.84 \cdot 10^{-4}$). Relaxation takes place until $t_{p,T}$ ($t^* = 2.79 \cdot 10^{-4}$).

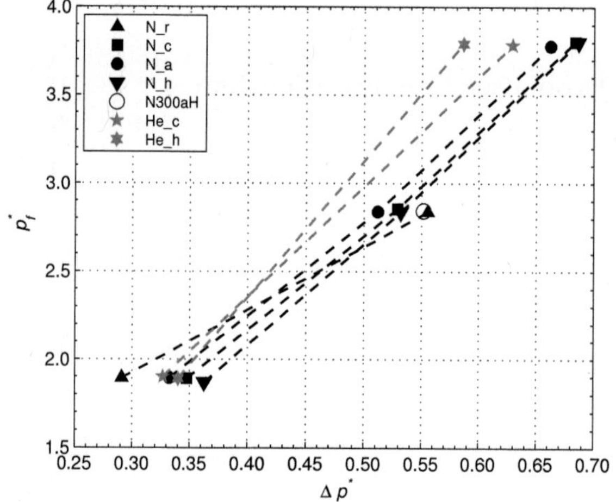

Fig. 6.25 Nondimensional maximal pressure drop Δp^* of the relaxation phase over the dimensionless final tank pressure p_f^* for all performed experiments; dashed lines are only for better visualization. N_r includes all data points for the N200r, N300r and N400r experiments and so on. All data can be found in Table A.8 in the Appendix.

phase. In section (A), the tank is pressurized up to 300 kPa. Afterwards, the pressurant gas mass flow is stopped and the tank is left closed during sections (B) and (C). After (C), the experiment is completed and the tank is opened again. It can nicely be seen, that the tank pressure drops remarkably after the pressurization end in section (B), but shows an asymptotic course to the end of phase (B). The maximal pressure drop between the pressurization end and the relaxation end is Δp. In section (C), the tank pressure rises again until the experiment is stopped. In Figure A.3 in the Appendix, the corresponding temperature evolutions are depicted.

In Figure 6.25, the dimensionless maximal pressure drop Δp^* over the dimensionless final tank pressure p_f^* is depicted for all performed experiments. The pressure drop Δp is defined as the absolute value of the difference between the tank pressure at the peak of the pressure curve (p_m) and the tank pressure at the relaxation end (p_T). All data are summarized in Table A.8 in the Appendix. Please note that the value of the tank pressure at the peak p_m does not always correspond to the final tank pressure p_f, as the valve of the pressurant gas inflow had to be closed manually, which results in a delay between p_f and p_m. Furthermore, for the nondimensionalization of the helium pressurized experiments in Figure 6.25, the nitrogen reference parameter are used to aid comparison. The nondimensional final tank pressures are summarized in Table A.8 in the Appendix.

It can be stated from Figure 6.25 that the pressure drop Δp^* is in general linearly dependent on the final tank pressure. The gradient of the linear dependency however, is different for each pressurant gas temperature and species (GN2 or GHe). Also the ullage volume has an influence on Δp as it can be seen by comparison of the N300a and N300aH experiments.

6.8.2 Analytical Determination

The analytical approach for the determination of the pressure drop was introduced in Section 2.6. Equation 2.66, which is used for the analytical determination of the pressure drop can be written for clearer presentation as

$$
p(t) = p_f \underbrace{\frac{\bar{T}_v(t)}{\bar{T}_{v,p,f}}}_{\text{int. energy}} - \underbrace{\frac{R_s \bar{T}_{v,p,f}\, \rho_{v,sat}(t)}{V_u} \left[\frac{2\, \dfrac{\rho_{l,sat}(t)}{\rho_{v,sat}(t)} \mathbf{Ja}(t) A_{cond}(t) \sqrt{D_{t,l}}}{\sqrt{\pi}} \left(\sqrt{t} - \sqrt{t_{p,f}} \right) \right]}_{\text{condensation}}
\tag{6.51}
$$

$$
+ \underbrace{\frac{R_s \bar{T}_{v,p,f}\, \rho_{v,sat}(t)}{V_u} \left[v_{evap}(t) A_{evap}(t) \left(t - t_{p,f} \right) \right]}_{\text{evaporation}}
$$

where $\mathbf{Ja}$, ρ and $\bar{T}_v$ are time dependent parameters. It can be seen that Equation 6.51 consists of one part which is pressure change due to the change in internal energy, one part which is pressure change due to condensation and one which is pressure change due

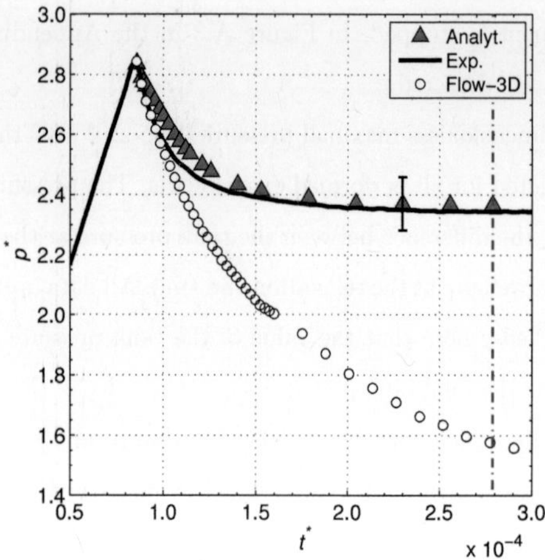

Fig. 6.26 N300h experiment: Experimental pressure drop, analytical pressure drop using Equation 6.51 and results of the Flow-3D simulations. The vertical line represents the time defined as relaxation end.

to evaporation. For the analysis of the experimental pressure drop with Equation 6.51, the evaporation velocity $v_{evap}(t)$, the condensation surface $A_{cond}(t)$ and the evaporation surface $A_{evap}(t)$ are not known. It is assumed that at the beginning of the relaxation phase condensation predominates and subsequently the amount of evaporated propellant increases. It is furthermore assumed that evaporation appears on the free surface on an annular area adjacent to the tank wall with increasing width. As the evolution over time of the width of this surface is not known, it is approximated for the analysis in this study that the evaporation and the condensation surface are half the area of the free surface over the whole relaxation phase $A_{cond}(t) = A_{evap}(t) = A_\Gamma/2$. The pressure drop due to evaporation is therefore probably overestimated and the pressure drop due to condensation underestimated. For the determination of the evaporation velocity $v_{evap}(t)$ the following assumption is made: At relaxation end it is supposed that the absolute values of the evaporation and condensation velocities are equal, as no pressure gradient occurs. On that account, the maximal value for the evaporation velocity $v_{evap}(t)$ in Equation 6.51 is chosen for the whole relaxation phase $|v_{evap}(t)| = |v_{cond}(t_{p,T})|$. Figure 6.26 depicts the results of Equation 6.51 with the applied assumptions, compared to the experimental pressure data for the N300h experiment. The initial vapor temperature used in Equation 6.51 is $T_{v,p,f} = 181$ K, the initial pressure for the pressure drop is $p_f = 300.1$ kPa, the time $t_{p,f} = 65.7$ s, the specific gas constant of ni-

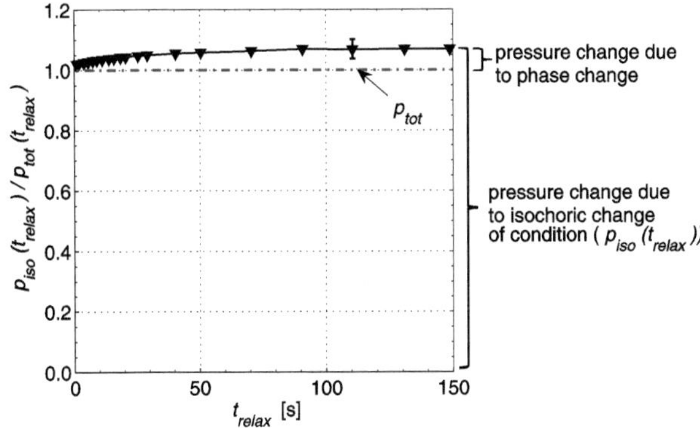

Fig. 6.27 N300h experiment: Percentage of pressure change due to phase change and pressure change due to isochoric change of condition $p_{iso}(t_{relax})$ of the total pressure change $p_{tot}(t_{relax})$ over the relaxation time t_{relax} ($t_{relax} = 0$ at pressurization end $t_{p,f}$).

trogen is $R_s = 296.8$ J/(kg K), the ullage volume $V_u = 0.014$ m^3 with the free surface area $A_\Gamma = 0.0688$ m^2, the thermal diffusivity is $D_{t,l} = 7.74 \cdot 10^{-8}$ m^2/s and the condensation velocity at relaxation end $v_{cond}(t_{p,T}) = -8.75 \cdot 10^{-5}$ m/s, calculated with Equation 2.62. In Figure 6.26, it can be seen that the evolution of the analytical pressure curve corresponds very well to that of the experiment over the whole relaxation time, which approves the approach of Equation 6.51 and the assumptions made. The vertical line in Figure 6.26 is the time defined as relaxation end ($t_{p,T}^* = 2.79$, see Table A.8 in the Appendix). The error bar, exemplarily for one data point, covers only the error due to the uncertainties of the determination of the mean vapor temperature $\bar{T}_v$. For this study, this was only feasible by averaging over the internal energy and assuming that the temperature distribution is linear between two temperature sensors (see Equation 6.13). Please note that Equation 6.51 represents a "smart fit" to the experimental data and does not allow an a priori calculation of the pressure drop as the time dependent parameters **Ja**, ρ and $\bar{T}_v$ have to be determined from experimental data.

In order to better understand the influence of the phase change and the change in internal energy over the relaxation phase, Figure 6.27 shows the ratios over the relaxation phase. The part of the pressure change in internal energy corresponds to an isochoric change of condition, depicted as $p_{iso}(t_{relax})$.

$$p_{iso}(t_{relax}) = p_f \frac{\bar{T}_v(t_{relax})}{\bar{T}_{v,p,f}} \tag{6.52}$$

This pressure is depicted at the rate of the pressure, calculated with the entire Equation 6.51 $p_{tot}(t_{relax})$. The relaxation time t_{relax} of Figure 6.27 is defined as $t_{relax} = t - t_{p,f}$. In Figure 6.27, it can be seen that the pressure change due to isochoric state of condition is the main contributor to the pressure drop. Pressure change by phase change is maximal 6 % of the total pressure change. This leads to the conclusion that the pressure drop mainly occurs due to the temperature decrease in the tank ullage and only to a smaller part due to phase change.

6.8.3 Numerical Results

The pressure drop is calculated with Flow-3D as a restart calculation from pressurization end. The mass source is therefore deactivated. Figure 6.26 also depicts the Flow-3D results of the pressure drop of the N300h experiment. It can be seen that a pressure drop occurs which fits very well to the experimental pressure evolution in the beginning of the pressure drop. It can therefore be stated that with the applied numerical model, the maximal gradient of the pressure drop is covered. Afterwards however, the pressure calculated by Flow-3D decreases much lower than that of the experiment. This is assumed to be due to the fact that the tank wall in the numerical model is only covered by one cell column, as the wall of the test tank has only a thickness of $1.5 \cdot 10^{-3}$ m. It might therefore result in a not very accurate resolution of the heat flows along the wall during the relaxation phase. As the relaxation phase is more than twice as long as the pressurization phase, the heat flows along the walls might significantly affect the pressure drop evolution. If the numerical pressure decrease is considered after the experimental relaxation end, also a horizontal pressure curve appears but at a much lower pressure and therefore later than in the experiments. Also the consideration of the ambient heat flow, presented in Section 4.4, does not greatly influence the numerical pressure drop. The Flow-3D results of the pressure drop of the helium pressurized experiments do not correspond well to the experimental results over the whole relaxation time. As already for the pressurization phase considerable deviations appear, it therefore results also in discrepancies between the numerical and the experimental data for the relaxation phase.

On that account, it can be summarized that the Flow-3D results of the pressure drop of the GN2 pressurized experiments can be used to determine the maximal pressure gradient right after pressurization end, analyses at a later time however should not be performed based on this data. The pressure drop simulations of the GHe pressurized experiments are not recommended for additional analyses. As the presented numerical model is designed with focus on the pressurization phase, other settings may be advantageous in order to further analyze the relaxation phase with Flow-3D. A finer resolution of the tank wall and therefore a more accurate consideration of the heat flow inside the tank wall might increase the accordance to the experimental pressure drop.

6.9 Overview of Main Factors

By means of the N300h experiment, an overview of the main factors of the active tank pressurization, described in the previous sections, is given. For this experiment, the initial tank pressure is 104.2 kPa and before pressurization start, the LN2 has an uniform temperature of 77.6 K and for the GN2 in the tank ullage the temperature increases linearly between the saturation temperature at the free surface of 77.6 K and the lid temperature of 278 K (Figure 6.28 (a)). The applied pressurant gas is GN2 with a mass flow of $8.32 \cdot 10^{-4}$ kg/s with a pressurant gas temperature of 352 K.

The final tank pressure of this experiment is 300 kPa and due to the injected gas mass, the pressure in the tank rises with a nearly linear slope (Figure 6.1 (a)). The resulting analytically determined maximal pressurant gas inlet velocities are between 2.68 m/s at 100 kPa and 0.89 m/s at 300 kPa, calculated by the pressurant gas mass flow and the diffuser outlet area. From the numerical simulations, maximal inlet velocities between 1.13 m/s at pressurization start and 0.49 m/s at pressurization end result. Figure 6.29 depicts the Flow-3D results for the pressurant gas velocity at the beginning of the pressurization and at pressurization end. It can be seen that for this case the pressurant gas flows along the lid and then the tank wall downwards. Please note that for the experiments with the lowest pressurant gas temperature of 144 K, the results of the Flow-3D simulations state that the injected gas does not reach the tank wall but flows after the injector directly downwards toward the free surface. The experimental pressurization time of the N300h experiment until the final tank pressure of 300 kPa is reached is 60.7 s (Flow-3D: 60.0 s). The resulting required pressurant gas mass is 0.0505 kg (Flow-3D: 0.0499 kg, see Table 6.1). At pressurization end, the vapor phase is still thermally stratified and the temperature gradient is greatly influenced by the pressurant gas temperature (see Figure 6.28 (b)). The analytical consideration of the heat transfer of the pressurization and relaxation phases shows that for the N300h experiment a change in heat for the pressurization period of $dQ_{0,f} = -6.784 \cdot 10^3$ J and for the re-

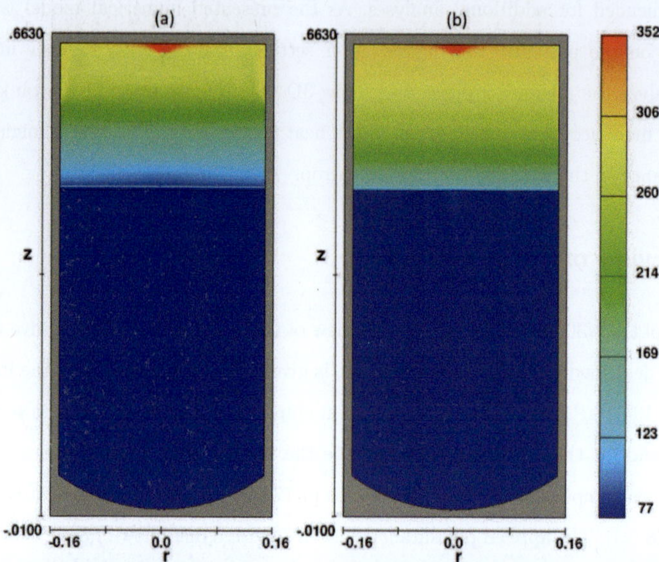

Fig. 6.28 N300h experiment: Fluid temperatures (in K) from Flow-3D at (a) 4 s after pressurization start and (b) at pressurization end.

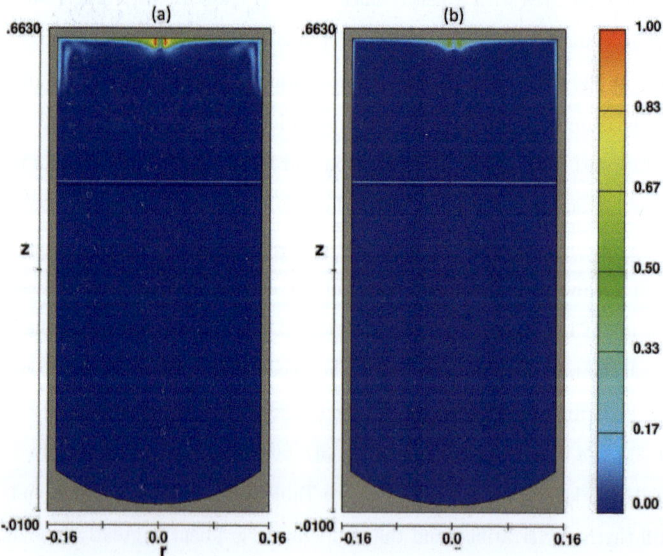

Fig. 6.29 N300h experiment: Velocity magnitude (in m/s) from Flow-3D at (a) 4 s after pressurization start and (b) at pressurization end.

laxation phase of $dQ_{f,T} = 3.608 \cdot 10^3$ J results. During pressurization, the pressurant gas with an enthalpy of $m_{pg}\,h_{pg} = 18.43 \cdot 10^3$ J is injected into the tank. This results in the fact that heat, $dQ_{v,0,f} = -11.116 \cdot 10^3$ J (Flow-3D: $dQ_{v,0,f} = -5.942 \cdot 10^3$ J), leaves the tank ullage toward the tank wall and a part of this heat, $dQ_{l,0,f} = 4.333 \cdot 10^3$ J (Flow-3D: $dQ_{l,0,f} = 1.167 \cdot 10^3$ J), is conducted downwards and enters the liquid just below the free surface (Figures 6.14 (a) and 6.19). During the relaxation period (Figure 6.16 (a)), only little energy is still transferred from the ullage to the tank wall ($dQ_{v,f,T} = -1.361 \cdot 10^3$ J), but the heat input into the liquid phase increases a little ($dQ_{l,f,T} = 4.969 \cdot 10^3$ J). Moreover, it is found that change in energy due to phase change plays only a small role on the overall consideration ($m_{cond,0,f}\,h_{v,cond,0,f} = 0.52 \cdot 10^3$ J, $m_{cond,f,T}\,h_{v,cond,f,T} = 0.55 \cdot 10^3$ J). Additionally to the heat input from the tank wall into the liquid phase, also the change in saturation temperature at the free surface due to the pressure rise contributes to a thermal stratification of the liquid. The resulting strong temperature gradient below the liquid surface at pressurization end decreases during relaxation due to heat conduction toward the bulk, which results in an increase of the thermal boundary layer thickness over time (see Section 6.3). During the pressurization phase, the boundary of the thermal stratified layer passes the temperature sensor T3, which is 0.005 m below the free surface and during relaxation, the thermal boundary layer increases beyond the temperature sensor T2, which is 0.015 m below the free surface (see Figure 6.1 (b) and Table A.1 in the Appendix). The analytical determination of the phase change mass states that during the pressurization phase 0.0064 kg GN2 condenses and during relaxation 0.0066 kg GN2 condenses (see Section 6.5.1.1). The Flow-3D result for the phase change during pressurization is 0.0066 kg, which corresponds well to the analytical value. During relaxation, 0.036 kg GN2 evaporates according to the numerical results, which is not consistent with the analytical value. The numerical pressure evolution during relaxation lies below the experimental pressure slope, even though evaporation is the main mode of phase change (see Figure 6.26 and Table 6.4). The maximal experimental pressure drop after pressurization end is 56.24 kPa and the experimental relaxation time is 152.3 s (see Figure 6.24). The main driver for the pressure drop is found to be the change in vapor temperature $\bar{T}_v(t)/\bar{T}_{v,p,f}$ (see Section 6.8.2), which has a minimum value of 0.89 at relaxation end for the N300h experiment. Only maximal 6 % of the pressure drop is due to phase change (see Figure 6.27). Please note that no steady state appears after relaxation end due to the heat input from the ambient (see Figure A.3 in the Appendix). As this study focuses on the pressurization and the relaxation phase, the subsequent pressure and temperature increase must be kept in mind, but is not focus of this study.

6.10 Upscaling of the Required Pressurant Gas Mass

Based on the scaling concept introduced in Section 2.7, a correlation between the nondimensional required pressurant gas mass and the two most relevant characteristic numbers for the active-pressurization the JAKOB number and the thermal expansion FROUDE number is found. The characteristic numbers are defined as follows:

$$\mathbf{Ja} = \frac{c_{p,l} \left(T_{sat,p,f} - T_{sat,p,0}\right)}{\Delta h_{v,p,f}}$$

$$\mathbf{Fr}_{\beta_T \mathbf{pg}} = \frac{v_{pg}^2}{g \ \beta_{T,pg} \left(T_{pg} - T_{sat,p,0}\right) \dfrac{V_u}{A_\Gamma}}$$

The following correlation between the nondimensional required pressurant gas mass $\tilde{m}_{pg,cn}$ and the JAKOB number and the thermal expansion FROUDE number is found

$$\tilde{m}_{pg,cn} = K \ \frac{\mathbf{Ja}^{3/2}}{\mathbf{Fr}_{\beta_T \mathbf{pg}}^{1/3}} \tag{6.53}$$

where K is a constant which includes the other dimensionless numbers, determined by dimensional analysis. The nondimensional required pressurant gas mass for the scaling $\tilde{m}_{pg,cn}$ is defined as $\tilde{m}_{pg,cn} = m_{pg}/m_{ref}$. In Equation 6.53, the JAKOB number considers the phase change part during pressurization and the thermal expansion FROUDE number considers the influence of the pressurant gas temperature and the aspect ratio of the propellant tank. Figure 6.30 shows the nondimensional required pressurant gas masses $\tilde{m}_{pg,cn}$ in comparison to the experimental nondimensional required pressurant gas masses $\tilde{m}_{pg,exp} = m_{pg,exp}/m_{ref}$ of the GN2 pressurized experiments of this study. It is found that the constant $K \approx 1$. With regard to the reference mass m_{ref}, very good results are obtained with $m_{ref} \approx 1$ kg. Table A.13 in the Appendix summarizes the relevant data and the results of the pressurant gas mass correlation for the experimental data of this study and shows that the correlation of Equation 6.53 does not hold for the helium pressurized experiments.

The pressurant gas mass correlation of Equation 6.53 is validated with data from literature. A study which provides all relevant data is the study of van Dresar and Stochl [34], who published in 1993 results of active-pressurization experiments of a flightweight LH2 tank, pressurized with GH2. The described tank is approximately an ellipsoidal volume with a

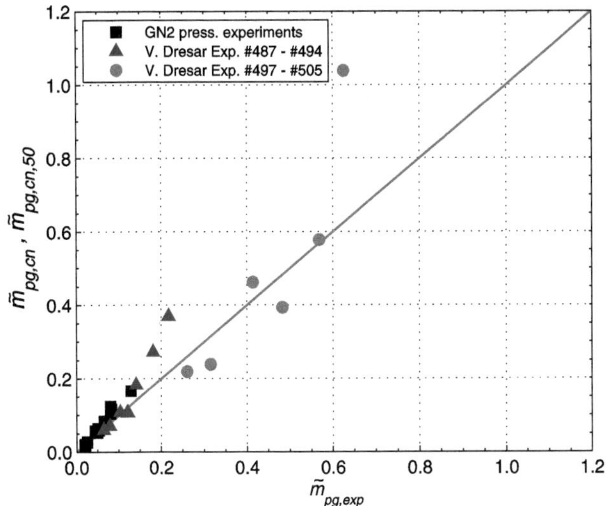

Fig. 6.30 Nondimensional experimental required pressurant gas masses for the scaling $\tilde{m}_{pg,cn}$ and $\tilde{m}_{pg,cn,50}$ over the nondimensional required pressurant gas masses $\tilde{m}_{pg,exp}$. Data points for GN2 pressurized experiments and experiments #487 - #494 by van Dresar and Stochl [34] are determined with Equation 6.53. Data points for experiments #497 - #505 by van Dresar and Stochl [34] are determined with Equation 6.54. The corresponding values can be found in Tables A.13 and A.14 in the Appendix.

total volume of 4.89 m^3. The ratio between major to minor axis is 1.2 with a major diameter of 2.2 m. The pressurant gas enters the tank via a diffuser with an outlet area of 0.057 m^2. In the presented study, two different ullage volumes and different pressurant gas inlet velocities are analyzed. Table A.14 in the Appendix summarizes the relevant data and the results of the applied scaling formula. Figure 6.30 shows $\tilde{m}_{pg,exp}$ over $\tilde{m}_{pg,cn}$ for the experiments #487 to #494 of van Dresar and Stochl, which have an ullage volume of approximately 13 %. It can be seen that the correlation provides quite good results for the required pressurant gas mass. For the experiments with the long pressurization times t_{press} however, the accuracy of the correlation reduces.

The experiments #497 to #505 of van Dresar and Stochl [34] have an ullage volume of approximately 50 % and, due to the ellipsoidal shape of the tank, a large free surface area of 3.8 m^2. For these experiments, it is found that the correlation presented in Equation 6.53 does not provide good results, as the free surface area is very big compared to the ullage volume. On that account, phase change has an increased influence on the required pressurant gas mass. It is found that the following modified form of Equation 6.53 provides quite accurate results for the required pressurant gas mass for the high aspect ratio

$$\tilde{m}_{pg,cn,50} = K \, \frac{\mathbf{Ja}}{\mathbf{Fr}_{\beta_T \mathbf{pg}}^{1/3}} \tag{6.54}$$

with $K \approx 1$ and $m_{ref} \approx 1$ kg. In Figure 6.30, it can be seen that this equation has good results for the required pressurant gas mass. Also for this correlation holds that the accuracy decreases with increasing pressurization time.

It can be summarized that the nondimensional required pressurant gas mass can be correlated with the JAKOB number and the thermal expansion FROUDE number. Based on the presented correlations the phase change, tank aspect ratio and pressurant gas temperature are the determining factors for the required pressurant gas mass. The form of the correlation is dependent on the aspect ratio between the ullage volume and the free surface area. This correlation is so far only applicable for evaporated-propellant pressurization systems and further analyses should be performed with regard to the reference mass m_{ref}.

Chapter 7
CONCLUSION AND DISCUSSION

Against the background of the new development of European cryogenic upper stages for the new launcher generation, interest in the advancement of cryogenic fluid management technologies has been significantly revived. One part of the cryogenic fluid management system is the tank pressurization system. The purpose of the pressurization system is to control and maintain the required pressure in the propellant tanks at any time of the mission. Therefore, a pressurant gas is injected into the propellant tank, which is usually evaporated propellant or an inert gas, stored in additional vessels. Due to the very low temperatures of cryogenic fluids, complex fluid-dynamic and thermodynamic processes occur during and after pressurization inside the propellant tank. It is not yet fully understood, which process occurs, when does it appear, and to what extent does it influence the pressurization process.

Since launcher tank pressurization requires the use of on-board fluids, optimization of the pressurization process is essential in order to lower the launcher mass and therefore to increase the payload mass. Proper predictions concerning the active-pressurization phase for cryogenic propellants for the launcher application require an improved understanding of the appearing complex fluid-dynamic and thermodynamic phenomena. The initial active-pressurization phase prior to engine ignition is therefore considered in this study, as it represents the preparatory phase which determines the initial conditions of the subsequent phases, such as engine ignition, propellant sloshing as well as propellant draining during engine operation phase. In this study, ground experiments were performed in a sub-scale propellant tank with liquid nitrogen as cryogenic model propellant, which was pressurized with gaseous nitrogen or gaseous helium. Numerical simulations were additionally performed, using the commercial CFD software Flow-3D and analytical approaches and correlations were applied.

Experimental research in the field of active-pressurization has been pursued mainly in the USA and primarily during the Apollo missions. Section 3.3 gives an overview of the experimental work, performed in the past. The majority of the experiments focuses on the pressurization phase in combination with propellant draining, which is relevant for the engine operation phases. The ground experiments, performed in this study however, focus on the initial active-pressurization phase without draining. In Chapter 4, the experimental setup with the subscale propellant tank is introduced. As pressurant gas, either gaseous nitrogen (GN2) or gaseous helium (GHe) is applied. The pressurant gas is fed with a constant mass flow into a heat exchanger. At this point, the pressurant gas temperature can be altered. Afterwards, the pressurant gas is injected via a diffuser into the experiment tank, which is partly filled with liquid nitrogen (LN2). Inside the experiment tank, the fluid and wall temperatures are measured at various positions. The tank pressure is also measured. For the presented experiments, the test tank is pressurized from ambient pressure to the final tank pressure. Three different final tank pressures are selected: 200 kPa, 300 kPa and 400 kPa. As pressurant gas temperatures, 144 K, 263 K, 294 K and 352 K are selected for the GN2 pressurization and 263 K and 352 K for the GHe pressurization. The test tank is placed in a vacuum container in order to reduce heat input from the ambient. It can be summarized that the applied test setup and the performed active-pressurization experiments give a good insight into the subject of active-pressurization of cryogenic propellants for the launcher application.

For the numerical considerations presented in this study, the commercial CFD program Flow-3D version 10.0 is used. Chapter 5 introduces the applied quasi 2-D time-dependent Flow-3D model of the experimental tank together with a sensitivity analysis of the mesh and the Flow-3D accommodation coefficient *rsize*. The Flow-3D accommodation coefficient *rsize* is a multiplier on the phase change rate. In the sensitivity study, no converging behavior is found for the mesh and/or the Flow-3D accommodation coefficient *rsize*. The numerical model is validated with data from the experimental pressure evolutions. For the simulation of the pressurization phase of the GN2 pressurized experiments, the applied numerical model with a cell size of 0.002x0.0025 m and a *rsize* value of 0.1 shows very good agreement with the experimental results. For the simulation of the experiments pressurized with GHe, a *rsize* value of 0.0001 has to be selected in order to enable numerical results, which leads to higher deviation from the experimental results. The relaxation phases of the experiments are simulated with the same numerical settings but without the mass source, which is used for the pressurization phase as substitute of the pressurant gas injector.

It can be summarized that the applied state-of-the art commercial CFD program Flow-3D is able to simulate the pressurization and relaxation phases of the performed active-pressurization experiments. The results of the simulations and whether results can be achieved are strongly dependent on the numerical model of the tank with the selected numerical settings, especially the Flow-3D accommodation coefficient *rsize* and the applied grid. It can be stated that good results are feasible with the Flow-3D model introduced in this study, which is designed with focus on the pressurization phase. In order to further analyze the relaxation phase with Flow-3D, other settings may be advantageous, such as the more accurate consideration of the heat flow inside the tank wall. Referring to the corresponding literature summarized in Section 3.2, studies from e.g. Hardy and Tomsik [51], Sasmal et al. [85], Adnani and Jennings [1], Wang et al. [99, 100, 101], Kwon et al. [58] and Leuva et al. [63] addressed the numerical consideration of the active-pressurization process. It can be stated however, that the numerical results presented in this study are the first published results with a commercial CFD program of the initial active-pressurization of cryogenic propellants in normal gravity with consideration of the liquid and the vapor phase and therefore also the mass transfer and the heat transfer over the free surface.

This study presents results of experimental, numerical and analytical analyses of the active-pressurization of cryogenic propellants. The following subjects are therefore addressed: pressure and temperature evolution, thermal stratification, required pressurant gas mass, phase change, heat transfer and the pressure drop after pressurization end.

The experimental results of the pressure and temperature evolution of the active-pressurization experiments are presented in Section 6.2. At initial conditions, the tank ullage is only filled with evaporated nitrogen, representing a two phase system with a single fluid. By injection of the pressurant gas, the tank pressure increases almost linearly up to the final pressure. In the analysis of the required pressurant gas mass of the experiments with respect to the pressurant gas temperature and species presented in Section 6.4.1, it is found that an increased pressurant gas temperature results in a steeper pressurization curve and therefore a faster pressurization. It results that the GN2 pressurized experiments with the highest pressurant gas temperature of 352 K needed the least pressurant gas mass for the pressure increase of the GN2 pressurized test cases. This is due to the fact that a higher specific heat capacity of the pressurant gas results in a faster tank pressure increase. Another interesting fact is that the tests with a pressurant gas temperature near the initial temperature of the inlet pipe require less pressurant gas mass than expected. This exposes that the heating respectively the cooling of the diffuser pipe from initial temperature to the pressurant gas temperature also increases the pressurant gas need.

By using gaseous helium as pressurant gas, the pressurization process is much faster. One reason for that is that helium has a higher specific heat capacity than the gaseous nitrogen (e.g. $c_{p,GHe} = 5.196{\cdot}10^3$ J/(kg K) and $c_{p,GN2} = 1.124{\cdot}10^3$ J/(kg K) at $p = 101.3$ kPa and $T = 77.35$ K [62]), resulting in an increased heat input into the vapor phase over the pressurization period, compared to the gaseous nitrogen pressurization, and therefore a faster pressure increase. Moreover, helium has a much lower density than nitrogen, which means that less GHe is needed to fill the tank ullage (e.g. vapor density for helium at $p = 106$ kPa and a mean ullage temperature of $T_v = 144$ K is 0.354 kg/m^3 and for nitrogen is 2.50 kg/m^3 [62]). Additionally, helium is a non-condensable gas and therefore cannot undergo phase change with the liquid nitrogen.

Based on the results of the analysis of the required pressurant gas masses, it can be stated that a pressurant gas with a very high specific heat capacity is recommended in order to reduce the pressurant gas requirements of cryogenic launcher stages. By exclusive consideration of the pressurant gas masses, the application of an evaporated propellant with the highest possible pressurant gas temperature or a non-condensable gas is therefore advantageous. Moreover, it confirmed to be relevant that the pressurization lines are already chilled down in advance, to reduce the required pressurant gas mass. These results agree with results from literature. Nein and Thompson [75] stated that the strongest influence on pressurant weight has the pressurant gas inlet temperature. Stochl et al. [90, 91, 92, 93] presented that an increase of the inlet gas temperature results in a decrease in the pressurant requirement for constant ramp rates and it was determined for all presented experiments that the inlet gas temperature has the strongest influence on the pressurant gas requirements. Van Dresar and Stochl [34] stated that the pressurant requirements increase with increasing ramp time and Wang et al. [101] affirms that a pressurant gas with a large specific heat capacity has a better pressurization performance.

The numerical results of the required pressurant gas masses of the experiments presented in Section 6.4.2 show that the results for the GN2 pressurized cases are in good agreement with the experimental results. Only for the experiments with the lowest pressurant gas temperature of 144 K, higher deviations appear for the final tank pressures of 200 kPa and 400 kPa due to discrepancies in the phase change mass flux, compared to the experimental data. For the GHe pressurized experiments, the Flow-3D accommodation coefficient *rsize* had to be decreased to 0.0001 to enable a stable simulation and therefore higher errors appear.

The analytical results of the phase change during active-pressurization, introduced in Section 6.5.1, are based on the experimental data and show that for the GN2 pressurized experiments condensation is the dominating mode of phase change during the pressurization phase. The lower the pressurant gas temperature, the longer the pressurization phase and the more GN2 condenses and cannot contribute to the pressure increase. During the pressurization with helium however, evaporation appears, which also facilitates a faster pressure increase for these experiments. The injected helium has still a quite high temperature when it reaches the free surface and therefore causes evaporation. For the experimental results of the short GN2 pressurization phases, evaporation or hardly any phase change is determined. This might be due to the fact that the warm GN2 causes first evaporation when reaching the free surface and then gets cooled down faster than the GHe, resulting in condensation for the longer pressurization phases. For the relaxation phase, condensation is determined for the experiments with GN2 and GHe as pressurant gases.

The numerical analysis of the phase change during pressurization shows that the numerical and the analytical determined phase change masses are in the same order of magnitude for the GN2 pressurized cases, which confirms the applied numerical model. For the helium pressurized cases, the Flow-3D results do not confirm the analytical results that evaporation predominates. These results however are not assumed to be very reliable due to the higher deviation from the experimental pressure curve because of the Flow-3D accommodation coefficient *rsize*. The numerical results of the phase change mass for the relaxation phase for both pressurant gas types are also not found to be plausible, as the overall pressure decrease in Flow-3D deviates strongly from the experimental course.

The achieved results for the phase change during the pressurization phase correspond well to statements from literature. Nein and Head [74] published that for the analyzed full-scale tanks, considerable mass transfer appears during expulsion between the gas in the tank ullage and the liquid propellant. The appearing mass transfer was always evaporation, except for the test, where the small model LN2 tank was pressurized with gaseous nitrogen, where condensation occurred during the entire discharge test. Nein and Thompson [75] stated that condensation occurred for all of the evaluated test cases except for one test with a high pressurant gas inlet temperature. Van Dresar and Stochl [34] presented experimental data of the pressurization and expulsion of a full-scale LH2 tank, pressurized with gaseous hydrogen. They stated that the mass transfer rate plays a significant role in ramp phases of practical duration. It was found that during a three minute ramp phase with a pressurant gas temperature of 280 K, mass transfer was not constant, but switches from evaporation to condensation and back to evaporation. Additionally it was stated that the pressurant requirements increase with increasing ramp time.

For the majority of the numerical studies in literature, phase change over the free surface is disregarded, e.g. in Sasmal et al. [85], Wang et al. [99, 100, 101], Kwon et al. [58] and Leuva et al. [63]. The numerical analysis of this study however reveals that for the initial active-pressurization phase and the relaxation phase, phase change can play a considerable role and should not be disregarded in the numerical calculations.

For a cryogenic launcher stage, accurate predictions of the appearing modes of phase change are relevant as heavy evaporation of the propellant might lead to an insufficient amount of propellant for the mission and if a lot pressurant gas condenses, the required pressure increase during the pressurized phases might not be reached.

Based on the pressure and temperature evolution of the experiments, the evolution of the thermal stratification in the fluids is analyzed in Section 6.3 over the pressurization and relaxation periods. In the experimental evolution of the vapor temperature during the pressurization phase, presented in Section 6.2, the influence of the pressurant gas temperature can be seen. The hot pressurant gas increases the temperature of the tank ullage with decreasing impact from the lid toward the free surface. After the end of the pressurization, the temperatures reduce analogously to the pressure. In the analysis of the thermal stratification of the vapor phase, it is found that in the presented test setup, the pressurant gas temperature is the main influencing factor on the thermal stratification of the vapor phase. The vapor phase of the presented experiments is almost linearly stratified at pressurization start between the saturation temperature at the free surface and the lid temperature. During pressurization, the temperature of the vapor phase increases, predominated by the pressurant gas temperature. During relaxation, the vapor temperature decreases again, on the one hand due to the ongoing heat transfer to the tank wall and on the other hand due to the homogenization of the vapor phase, attempting to reach a quasi-steady state. The analytical approach of steady-state heat conduction in a flat plate is applied with good accordance to the experimental data.

For the liquid phase, no thermal stratification appears in the presented experiments at pressurization start, due to the preceding boil-off phase. At pressurization end however, the temperature of the free surface, which is assumed to correspond to the saturation temperature of the current tank pressure, is increased due to the pressure increase. This results in a strong temperature gradient, but only in the upmost liquid layers. In the relaxation phase, the tank pressure decreases, resulting in a decrease in the temperature of the free surface and the thickness of the thermal stratified liquid layer increases. The vertical temperature evolution has now a lower gradient but a higher penetration depth. The vertical temperature profiles at pressurization end and at relaxation end can be described quite accurately by the analytical model of transient heat transfer in large media with constant initial temperature. This approach was also used by Clark [24], to describe the heat transfer during pressurization from the free surface of a condensate film, appearing on a floating styrofoam piston, through the piston itself. It can be concluded that one part of the temperature increase in the liquid phase during pressurization and relaxation is due to the change in tank pressure and the corresponding vertical heat conduction, resulting in an increasing thickness of the thermal stratified layer over time.

The proper determination of the amount of heated propellant in the thermal stratified layers is important for a cryogenic upper stage, as the maximal allowed propellant temperature in the turbopumps may not be exceeded. The propellant, which is too warm for the turbopumps must not be used to generate thrust and remains in the propellant tanks as thermal residuals.

After reaching the final tank pressure in the experiments, the pressurant gas inflow is stopped and the tank is left closed. In the subsequent relaxation phase, the tank pressure decreases rapidly and the pressure curve shows an asymptotical evolution. The position where the pressure curve shows a horizontal course is defined as relaxation end. The analysis of the pressure drop (Section 6.8) leads to the awareness that for a constant tank pressure in a cryogenic propellant tank, continuous pressurization is essential. The total amount of the pressure drop depends on the final tank pressure of the pressurization phase, the pressurant gas temperature and the fluid, applied as pressurant gas. It is found that during this relaxation phase, only condensation appears, independently of the used pressurant gas. A "smart fit" equation is presented, which describes the evolution of the pressure drop for the performed experiments quite accurately (see Section 6.8.2). This approach cannot be applied a priori, however it enables the identification of the dominating parts, contributing to the pressure drop. It is found that the pressure drop after pressurization end is mainly driven by the change in vapor temperature and only to a small part due to condensation. It therefore follows that by reducing the heat transfer from the vapor phase to the tank wall, the pressure drop after pressurization end can be reduced. The Flow-3D pressure curve of the relaxation phase is in good agreement with the experimental data at the beginning of the relaxation phase, covering the maximal gradient of the pressure drop quite accurately. Also the overall course is similar, the tank pressure at relaxation end however is considerably lower than that of the experiments. This is assumed to be due to the fact that the applied Flow-3D model is designed with focus on the pressurization phase and therefore the modeling of the heat transfer through the tank wall during the relaxation phase might not be accurate enough to proper simulate the pressure drop. The results of the analysis of the pressure drop after pressurization end correspond to statements from Leuva et al. [63]. They presented that a pressure drop appears when no pressurant gas is injected and that a decreased temperature of the tank wall increases the pressure drop.

The experimental temperature evolution of the tank wall in Section 6.2 shows that it is affected by the pressurization process, but a much smaller temperature increase appears than for the vapor phase due to the slow reaction of the wall material. During the relaxation phase, the wall temperature shows hardly any changes. The analysis of the heat transfer in Section 6.6 shows that the dominating way of heat transfer during the pressurization phase is the heat transfer from the pressurant gas to the tank wall. The amount of heat transfer is therefore dependent on the pressurant gas temperature. This confirms results from literature. With regard to the heat transfer, van Dresar and Stochl [34] stated in 1993 that the largest portion of the input energy goes into wall heating. In 1962, Nein and Head [74] published that for a full-scale Saturn liquid oxygen system and a large single LOX tank, the major internal heat transfer takes place between the ullage gas and the liquid surface and that the heat transfer between the ullage gas and the adjacent tank walls are of minor importance. However, the results for the small model tank indicated a reversal in the predominant mode of heat transfer. It is moreover found in this study that a part of the heat, which gets into the tank wall during the pressurization phase in the area of the tank ullage, is conducted downwards and enters the liquid phase right below the free surface, increasing the temperature in the area nearby the tank wall. This heat input also contributes to the increase of the thickness of the thermal boundary layer below liquid surface over time. The heat input into the liquid phase also continues during the relaxation phase as there is still a considerable temperature difference to the liquid temperature.

Based on the results of the boil-off experiment introduced in Section 4.4, the total heat flow from ambient into the test tank is 35.5 W. For the numerical analyses, it is found that this heat flow can be disregarded for the pressurization phase of the analyzed tank, as the heat input by the pressurant gas is much stronger. During the numerical consideration of the relaxation phase, the heat input from ambient is also found to be negligible. However, for the design of a cryogenic upper stage throughout a whole mission, the heat transfer from ambient into the propellant tanks must not be disregarded. The amount of heat input from the outside becomes considerable especially during the ascent phase due to the aerothermal loads as well as the solar radiation for mission phases above the Earth's atmosphere.

By means of the wall to fluid heat flux calculated with Flow-3D, which is presented in Section 6.6.2, it can be stated that for the pressurization phase of the applied experimental setup a laminar forced convection predominates in the vapor phase and that in the uppermost liquid layers a laminar free convection flow dominates for the analyzed test case. Adnani [1] also stated the heat transfer in the dome region is primarily driven by forced convection and that it is a strong function of the diffuser design. This has to be kept in mind in the early stages of the development of a cryogenic launcher stage concept.

In Section 6.7, a correlation is derived for the presented experiments and test setup, which can be used for an a priori assessment of the pressure rise during pressurization. The correlation can be used for the GHe pressurized test cases and, in a more simplified form, for the GN2 pressurized experiments. It is based on the ideal gas law and the first law of thermodynamics for open systems. Furthermore, phase change is disregarded and as mode of heat transfer, only forced convection between the pressurant gas and the tank wall is regarded. For pressurization with helium, an ideal gas mixture in the tank ullage is assumed. The results are compared to experimental results with a very good accordance. Only for the N400r experiment, it became clear that for such low pressurant gas temperatures, phase change cannot be disregarded in determination of the pressure rise. For the GHe pressurized test cases, the partial pressure of helium has to be presumed, which makes it challenging to obtain very accurate results.

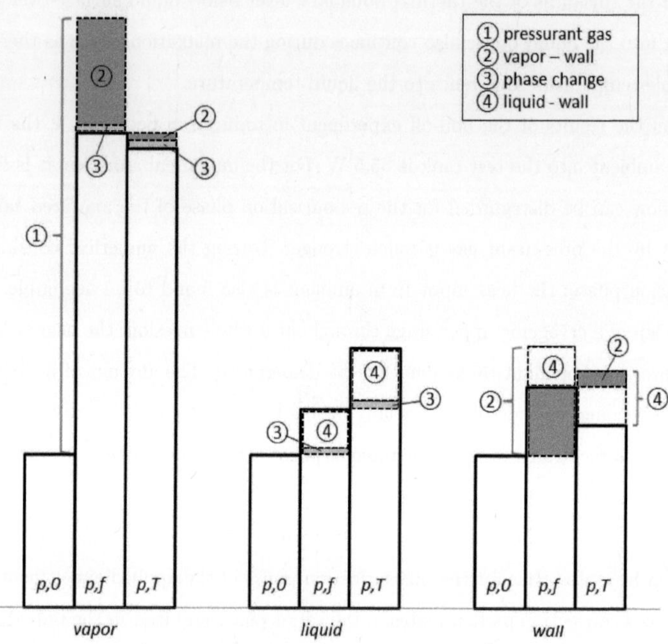

Fig. 7.1 Schematical energies of the N300h experiment of the vapor phase, the liquid phase and the tank wall at pressurization start $(p, 0)$, pressurization end (p, f) and at relaxation end (p, T). Considered changes between the inital state and pressurization respecitvely relaxation end are due to the injected pressurant gas (1), heat transfer between the vapor phase and the tank wall (2), heat transfer due to phase change (3) and heat transfer between the liquid phase and the tank wall (4).

In this study, it is found that the required pressurant gas mass for the active-pressurization correlates with the JAKOB number **Ja** of the liquid propellant and the thermal expansion FROUDE number of the pressurant gas $\mathbf{Fr}_{\beta_T \mathbf{pg}}$. The correlations, presented in Section 6.10 fit well for the GN2 pressurized experiments of this study and are validated with data from the study of van Dresar [34], who used GH2 to pressurize LH2. Based on the presented correlations, the phase change, tank aspect ratio and pressurant gas temperature are identified as determining factors for the required pressurant gas mass. In a next step, it might be possible to extend the correlations to be also applicable for helium pressurization.

Figure 7.1 depicts schematically the energy of the vapor phase, the liquid phase and the tank wall at pressurization start $(p, 0)$, pressurization end (p, f) and at relaxation end (p, T)

using the example of the GN2 pressurized experiment N300h. The change in energy of the vapor phase during the pressurization period is on the one hand due to the specific enthalpy and mass of the injected pressurant gas, which enters the control volume of the vapor phase (compare Figure 6.13). On the other hand, energy leaves due to heat transfer to the tank wall and phase change, here depicted in the form of condensation. During the relaxation phase, the energy of the vapor system decreases further due to the ongoing heat transfer and phase change. The energy change of the liquid system during the pressurization and the relaxation phase is an increase in energy due to the phase change and heat input from the tank wall. During the pressurization phase, the tank wall absorbs energy from the vapor phase, but it also releases energy into the liquid phase. The overall energy of the tank wall however increases. During the relaxation phase, the energy of the wall decreases, as the amount of heat transfer into the liquid phase is bigger than the heat input from the vapor phase. This representation of the major mechanisms of change in energy provides a good understanding of the effects during and after active-pressurization of cryogenic propellants, which is essential for the design and dimensioning of a cryogenic propellant tank and pressurization system. Please note that in a cryogenic propellant tank no steady-state exists as due to the low fluid temperatures continuous heat and mass transfer occurs, resulting in ongoing pressure and temperature changes (compare Figure A.3 in the Appendix).

Chapter 8
OUTLOOK

In the analysis of the experimental pressures curve during pressurization and relaxation it became clear that in order to provide a constant tank pressure, continuous pressurization is required, also without propellant draining. As the change in vapor temperature is the main driver of the pressure drop, adequate design of the tank wall thickness and material as well as the insulation concept is advantageous in order to reduce this effect. In order to decrease the required pressurant gas mass, a gas with a very high specific heat capacity is advantageous. It has to be kept in mind however, that e.g. the use of a heat exchanger for the increase in pressurant gas temperature has to be considered in the mass budget. Constraints for the maximum applicable pressurant gas temperature are determined by component reliability issues of the pressurization system, the tank wall temperature limitations and the effort to minimize thermal residuals. For the non-condensable pressurant gas, additional vessels have to be taken on board to store the pressurant gas. This results in an additional increase of the stage's mass.

The analysis of the heat transfer made clear that during the pressurization phase, the pressurant gas is the source of the major heat input. A considerable part of that heat goes into the tank walls and does not contribute to the pressure increase. The warm tank wall then heats the liquid up, which means that an increased pressurant gas temperature also results in an increased heat transfer to the tank wall and an increased liquid temperature. This has to be considered in the process of the design of the tank wall thickness, material and the insulation concept of a cryogenic launcher stage. During the pressurization phase with GN2, condensation appears, which results in a lower pressure increase. The increase in liquid propellant temperature has to be assessed as the turbopumps of the engine require the liquid propellants at a certain temperature to guarantee constant thrust. During pressurization with helium, evaporation predominates the phase change. The advantage is that the tank pressure increases faster, the disadvantage is that this amount of evaporated liquid propellant cannot be used for the engine to generate thrust without further equipment, e.g. a reliquefier.

This study also reaffirms that the design of the pressurant gas diffuser is very important as it determines the heat transfer to the tank or dome wall. As it was already found by Adnani [1], the heat transfer in the tank dome region is mainly driven by forced convection and it is a strong function of the diffuser design. Kendle [56] stated that diffuser-type pressurant injectors reduce the disturbance of the vapor phase. The gas inflow, if it is not well diffused, can cause mixing of the surrounding ullage, disrupting the thermal stratification and therefore increasing the gas-wall heat-transfer rate. Wang et al. [100] confirms that the use of a straight pipe injector can reduce the pressurant gas requirements but results in significant evaporation of the liquid propellant. This leads to the conclusion that a careful trade-off between the pressurant gas striking the liquid surface or the tank or dome wall and the disturbance of the vapor stratification has to be made.

References

1. P. Adnani and R.W. Jennings, Pressurization Analysis of Cryogenic Propulsion Systems: AIAA 2000-3788, in 36th AIAA/ASME/SAE/ASEE Joint Propulsion Conference & Exhibit, Huntsville, Alabama, USA, 2000.

2. V. Ahuja, A. Hosangadi, C.P. Lee, R.E. Field and H. Ryan, Computational Analyses of Pressurization in Cryogenic Tanks, AIAA 2008-4752, in 44th AIAA/ASME/SAE/ASEE Joint Propulsion Conference & Exhibit, Hartford, Connecticut, USA, 2008.

3. T. Arndt, Sloshing of Cryogenic Liquids in a Cylindrical Tank under Normal Gravity Conditions, PhD Thesis, 1. ed., Cuvillier Verlag, Göttingen, 2012.

4. R.W. Arnett and D.R. Millhiser, A Method for Analyzing Thermal Stratification and Self-Pressurization in a Fluid Container, NBS report 8777, NBS project 31505-40-3150451, 1965.

5. R.W. Arnett and R.O. Voth, A Computer Program for the Calculation of Thermal Stratification and Self-Pressurization in a Liquid Hydrogen tank, NASA-CR-2026, 1972.

6. V.S. Arpaci and J.A. Clark, Dynamic Response of Fluid and Wall Temperatures during Pressurized Discharge for Simultaneous, Time-Dependent Inlet Gas Temperature, Ambient Temperature and/or Ambient Heat Flux, in Advances in Cryogenic Engineering, vol. 7, Plenum Press, New York (1962), 419–432.

7. V.S. Arpaci, J.A. Clark and W.O. Wiener, Dynamic Response of Fluid and Wall Temperatures During the Pressurized Discharge of a Liquid from a Container, in Advances in Cryogenic Engineering, vol. 6, Plenum Press, New York (1961), 310–322.

8. J.C. Aydelott, Self Pressurization of Liquid Hydrogen Tankage, Master Thesis, Cornell University, 1976.

9. H.D. Baehr and K. Stephan, Wärme- und Stoffübertragung, 6. ed., Springer, Berlin, 2008.

10. D.O. Barnett, T.W. Winstead and L.S. McReynolds, An Investigation of LH2 Stratification in a Large Cylindrical Tank of the Saturn Configuration, in Advances in Cryogenic Engineering, vol. 10, Plenum Press, New York (1964), 314–324.

11. R.F. Barron, Cryogenic Heat Transfer, Series in chemical and mechanical engineering, Taylor and Francis, Philadelphia and Pennsylvania, 1999.

12. S. Barsi and M. Kassemi, Numerical and Experimental Comparisons of the Self-Pressurization Behavior of an LH2 Tank in Normal Gravity, Cryogenics (48), 2008, 122–129.

13. R.B. Bird, W.E. Stewart and E.N. Lightfoot, Transport Phenomena, Wiley/VCH, Weinheim, 1966.

14. F. Bizjak, W.J. Hines, C. OKI and D.J. Simkin, Analysis of Stored Gas Pressurization Systems for Propellant Transfer, Journal of Spacecraft and Rockets 1 (1), 1964, 103–107.

15. D.C. Bowersock, R.W. Gardner and R.C. Reid, Pressurized Transfer of Cryogenic Liquids, in Advances in Cryogenic Engineering, vol. 4, Plenum Press, New York (1960), 342–356.

16. D.C. Bowersock and R.C. Reid, An Analytical Method for Estimating Gas Requirements in the Pressurization and Transfer of Cryogenic Fluids, in Advances in Cryogenic Engineering, vol. 4, Plenum Press, New York (1960), 261–271.

17. Bronkhorst High-Tech BV, FLUIDAT on the Net: V1.38/6.18: Online Mass Flow and Physical Properties Calculation Tool, 20.12.2012.

18. Van P. Carey, Liquid-Vapor Phase-Change Phenomena: An Introduction to the Thermophysics of Vaporization and Condensation Processes in Heat Transfer Equipment, Taylor & Francis, 1992.

19. Y.A. Cengel, A.J. Ghajar and D. Pitts, Heat & Mass Transfer + EES DVD + Schaum's Outline of Heat Transfer, 4th ed., McGraw-Hill Science/Engineering/Math, 2010.

20. J.H. Chin, J.O. Donaldson, L.W. Gallagher, E.Y. Harper, S.E. Hurd and H.M. Satterlee, Analytical and Experimental Study of Liquid Orientation and Stratification in Standard and Reduced Gravity Fields, 2-05-64-1, Lockheed Missiles & Space Company, Sunnyvale, California, 1964.

21. J.H. Chin, E.Y. Harper, F.L. Hines, S.E. Hurd and G.C. Vliet, Analytical and Experimental Study of Liquid Orientation and Stratification and Liquid Ullage Coupling, 2-05-56-01, Lockheed Missiles & Space Company, Sunnyvale, California, 1965.

22. J.A. Clark, A Review of Pressurization, Stratification, and Interfacial Phenomena, in International Advances in Cryogenic Engineering, vol. 10, Plenum Press, New York (1965), 259–283.

23. J.A. Clark, G.J. van Wylen and S.K. Fenster, Transient Phenomena Associated with the Pressurized Discharge of a Cryogenic Liquid from a Closed Vessel, in Advances in Cryogenic Engineering, vol. 5, Plenum Press, New York (1960), 467–480.

24. J.A. Clark, W.A. Warren and H. JR. Merte, Pressurization of Liquid Oxygen Containers, Report 2646-9-P, Engineering Research Institute, University of Michigan, 1958.

25. E.F. Coxe and J.W. Tatom, Analysis of the Pressurizing Gas Requirements for an Evaporated Propellant Pressurization System, in Advances in Cryogenic Engineering, vol. 7, Plenum Press, New York, 234–243.

26. Crysler Corporation, Space Division, Evaluation of AS-203 Low Gravity Orbital Experiments, BB-3.4.3-5-101, 1967.

27. S.P. Das and E.J. Hopfinger, Mass Transfer Enhancement by Gravity Waves at a Liquid-vapour Interface, International Journal of Heat and Mass Transfer 52 (5-6), 2009, 1400–1411.

28. R.L. DeWitt and T.O. McIntire, Pressurant Requirements for Discharge of Liquid Methane from a 1.52-Meter (5-ft-) Diameter Spherical Tank under Both Static and Slosh Conditions, NASA TN D-7638, 1974.

29. R.L. DeWitt, R.J. Stochl and W.R. Johnson, Experimental Evaluation of Pressurant Gas Injectors during the Pressurized Discharge of Liquid Hydrogen, NASA TN D-3458, 1966.

30. N.T. van Dresar, Pressurization of Cryogens: A Review of Current Technology and its Applicability to Low-Gravity Conditions, AIAA-92-3061, 1992.

31. N.T. van Dresar, Prediction of Pressurant Mass Requirements for Axisymmetric Liquid Hydrogen Tanks, Journal of Propulsion and Power 13 (6), 1997, 796–799.

32. N.T. van Dresar, PVT Gauging with Liquid Nitrogen, Cryogenics, 46 (2-3), 2006, 118–125.

33. N.T. van Dresar, C.S. Lin and M.M. Hasan, Self-Pressurization of a Flightweight Liquid Hydrogen Tank - Effects of Fill Level at Low Wall Heat Flux, NASA Technical Memorandum 105411, AIAA-92-0818, in 30th Aerospace Science Meeting and Exhibit, Reno, Nevada, USA, 1992.

34. N.T. van Dresar and R.J. Stochl, Pressurization and Expulsion of a Flightweight Liquid Hydrogen Tank, in 29th Joint Propulsion Conference and Exhibit, Monterey, California, USA, AIAA-93-1966, 1993, NASA Technical Memorandum 106427, 1993.

35. F.A.L. Dullien, Porous Media Fluid Transport and Pore Structure, 1st edition, Academic Press, New York, 1979.

36. E. Dumont, H. Burkhardt, M. Sippel, M. Johannsson, C. Ludwig and E. David, Exploiting Technological Synergies for Future Launch Vehicles, in 4th IAA Conference – Space Technologies: Present and Future, Dnepropetrowsk, Ukraine, 2013.

37. E. Dumont, C. Ludwig, A. Kopp and M. Sippel, Advances TSTO Launch Vehicles, in 4th European Conference for Aerospace Sciences (EUCASS), Saint Petersburg, Russia, 2011.

38. M. Epstein, Prediction of Liquid Hydrogen and Oxygen Pressurant Requirements, in International Advances in Cryogenic Engineering, vol. 7, Plenum Press, New York (1965), 303–307.

39. M. Epstein and R.E. Anderson, An Equation for the Prediction of Cryogenic Pressurant Requirements for Axisymmetric Propellant Tanks, in Advances in Cryogenic Engineering, vol. 13, Plenum Press, New York (1968), 207–214.

40. M. Epstein, H.K. Georgius and R.E. Anderson, A Generalized Propellant Tank-pressurization Analysis, in International Advances in Cryogenic Engineering, vol. 10, Plenum Press, New York (1964), 290–302.

41. ESA, Proposal for an Adapted Ariane 5 ME and Proposal for Ariane 6, http://spaceinimages.esa.int/var/esa/storage/images/esa_multimedia/images/2012/11/proposal_for_an_adapted_ariane_5_me_and_proposal_for_ariane_6/12111904-7-eng-GB/Proposal_for_an_Adapted_Ariane_5_ME_and_proposal_for_Ariane_6.jpg, 12.03.2013.

42. S.K. Fenster, G.J. van Wylen and J.A. Clark, Transient Phenomena Associated with the Pressurization of Liquid Nitrogen Boiling at Constant Heat Flux, in Advances in Cryogenic Engineering, vol. 5, Plenum Press, New York (1960), 226–234.

43. Festo AG & Co. KG, http://www.festo.com/cat/de_de/data/doc_de/PDF/DE/SILENCERS _DE.PDF, 18.10.2013.

44. Flow Science, Inc., FLOW-3D User Manual -Version 10.0, 2011.

45. A. van Foreest, Modeling of Cryogenic Sloshing Including Heat and Mass Transfer, AIAA-2010-6891, in 46th AIAA/ASME/SAE/ASEE Joint Propulsion Conference & Exhibit, Nashville, Tennessee, USA, 2010.

46. D.F. Gluck and J.F. Kline, Gas Requirements in Pressurized Transfer of Liquid Hydrogen, in Advances in Cryogenic Engineering, vol. 7, Plenum Press, New York (1962), 219–233.

47. G.D. Grayson, Coupled Thermodynamic-Fluid-Dynamic Solution for a Liquid-Hydrogen Tank, in Journal of Spacecraft and Rockets 32 (5), 1995, 918–921.

48. G.D. Grayson, A. Lopez, F.O. Chandler, L.J. Hastings, A. Hedayat and J. Brethour, CFD Modeling of Helium Pressurant Effects on Cryogenic Tank Pressure Rise Rates in Normal Gravity, AIAA 2007-5524, in AIAA Joint Propulsion Conference and Exhibit, Cincinnati, Ohio, USA, 2007.

49. G.D. Grayson, A. Lopez, F.O. Chandler, L.J. Hastings and S.P. Tucker, Cryogenic Tank Modeling for the Saturn AS-203 Experiment, AIAA 2006-5258, in 42nd AIAA/ASME/SAE/ASEE Joint Propulsion Conference and Exhibit, Sacramento, California, USA, 2006.

50. G.D. Grayson and J. Navickas, Interaction between Fluid-dynamic and Thermodynamic Phenomena in a Cryogenic Upper Stage, AIAA-93-2753, in AIAA 28th Thermophysics Conference, Orlando, Florida, USA, 1993.

51. T.L. Hardy and T.M. Tomsik, Prediction of the Ullage Gas Thermal Stratification in a NASP Vehicle Propellant Tank Experimental Simulation using FLOW-3D: NASA Technical Memorandum 103217, 1990.

52. D.J.E Harvie, M.R Davidson and M. Rudman, An Analysis of Parasitic Current Generation in Volume of Fluid Simulations. Applied Mathematical Modelling, 30 (10), 2006, 1056–1066.

53. M.M. Hasan, C.S. Lin and N.T. van Dresar, Self-Pressurization of a Flightweight Liquid Hydrogen Storage Tank Subjected to Low Heat Flux, NASA Technical Memorandum 103804, in ASME/AIChE National Heat Transfer Conference, Minneapolis, Minnesota, USA, 1991.

54. E.J. Hopfinger and S.P. Das, Mass Transfer Enhancement by Capillary Waves at a Liquid–Vapour Interface, Experiments in Fluids, 46 (4), 2009, 597–605.

55. J.C. Humphrey, Pressurized Transfer of Cryogenic Fluids from Tanks in Liquid-Nitrogen Baths, in Advances in Cryogenic Engineering, vol. 6, Plenum Press, New York (1961), 281–292.

56. D.W. Kendle, Ullage Mixing Effects on Tank Pressurization Performance, Journal of Spacecraft and Rockets 8 (9), 1971, 990–992.

57. K.H. Kim, H.-J. Ko, K. Kim, Y.-S. Jung, S.-H. Oh and K.-J. Cho, Transient Thermal Analysis of a Cryogenic Oxidizer Tank in the Liquid Rocket Propulsion System during the Prelaunch Helium Gas Pressurization, Journal of Engineering Thermophysics 21 (1), 2012, 1–15.

58. O. Kwon, B. Kim, G. Kil, I. Cho and Y. Ko, Modeling the Prediction of Helium Mass Requirement for Propellant Tank Pressurization, in Journal of Spacecraft and Rockets 49 (6), 2012, 1150–1158.

59. J. Lacapere, B. Vieille and B. Legrand, Experimental and Numerical Results of Sloshing with Cryogenic Fluids, in 2nd European Conference for Aerospace Sciences (EUCASS), Brussels, Belgium, 2007.

60. R.F. Lacovic, Comparison of Experimental and Calculated Helium Requirements for Pressurization of a Centaur Liquid Oxygen Tank, NASA TM X-2013, 1970.

61. B. Leitenberger, Europäische Trägerraketen Band 1: Von der Diamant zur Ariane-4 - Europas steiniger Weg in den Orbit, 1. ed., Books on Demand, Norderstedt, 2009.

62. E.W. Lemmon, M.L. Huber and M.O. McLinden, NIST Standard Reference Database 23: Reference Fluid Thermodynamic and Transport Properties-REFPROP. NIST Standard Reference Database Number: 23 Version 9.0, 2010.

63. D. Leuva, S. Gangadharan, P. Wilson and B. Kutter, A CFD Study of Cryogenic LH2 Tank Ullage Pressurization, AIAA 2012-1888, 53rd AIAA/ASME/ASCE/AHS/ASC Structures, Structural Dynamics and Materials Conference, Honolulu, Hawaii, USA, 2012.

64. A. Lopez, G.D. Grayson, F.O. Chandler, L.J. Hastings and A. Hedayat, Cryogenic Pressure Control Modeling for Ellipsoidal Space Tanks, AIAA 2007-5552, July 2007.

65. C. Ludwig and M.E. Dreyer., Analyses of Cryogenic Propellant Tank Pressurization based upon Ground Experiments, AIAA 2012-5199, in AIAA Space 2012 Conference & Exhibit, Pasadena, California, USA, 2012.

66. C. Ludwig, M.E. Dreyer and E.J. Hopfinger, Pressure Variations in a Cryogenic Liquid Storage Tank Subjected to Periodic Excitations, International Journal of Heat and Mass Transfer 66, 2013, 223–234.

67. P.A. Masters, Computer Programs for Pressurization (RAMP) and Pressurized Expulsion from a Cryogenic Liquid Propellant Tank, NASA TN-D-7504, 1974.

68. S.J. Mattick, C.P. Lee, A. Hosangadi and V. Ahuja, Progress in Modeling Pressurization in Propellant Tanks, AIAA 2010-6560, in 46th AIAA/ASME/SAE/ASEE Joint Propulsion Conference & Exhibit, Nashville, Tennessee, USA, 2010.

69. A.M. Momenthy, Propellant Tank Pressurization-System Analysis, in Advances in Cryogenic Engineering, vol. 9, Plenum Press, New York (1964), 273–283.

70. R.W. Moore, A.A. Fowle, B.M. Bailey, F.E. Ruccia and R.C. Reid, Gas-Pressurized Transfer of Liquid Hydrogen, in Advances in Cryogenic Engineering, vol. 5, Plenum Press, New York (1960), 450–459.

71. M.E. Moran, N.B. McNelis, M.T. Kudlac, M.S. Haberbusch and G.A. Satorino, Experimental Results of Hydrogen Slosh in a 62 Cubic Foot (1750 Liter) Tank, NASA Technical Memorandum 106625, AIAA-94-3259, 1994.

72. National Aeronautics and Space Administration, Pressurization Systems for Liquid Rockets, NASA SP-8112, NASA Space Vehicle Design Criteria (Chemical Propulsion), NASA, Washington DC, 1975.

73. J. Navickas, Prediction of a Liquid Tank Thermal Stratification by a Finite Difference Computing Method, AIAA-88-2917, in AIAA/ASME/SAE/ASEE 24th Joint Propulsion Conference, Boston, Massachusetts, USA, 1988.

74. M.E. Nein and R.R. Head, Experiences with Pressurized Discharge of Liquid Oxygen from Large Flight Vehicle Propellant Tanks, in Advances in Cryogenic Engineering, vol. 7, Plenum Press, New York (1962), 244–250.

75. M.E. Nein and J.F. Thompson, Experimental and Analytical Studies of Cryogenic Propellant Tank Pressurant Requirements, NASA TN D-3177, 1966.

76. W.A. Olsen, Experimental and Analytical Investigation of Interfacial Heat and Mass Transfer in a Pressurized Tank Containing Liquid Hydrogen, NASA TN-3219, 1966.

77. W. Polifke and J. Kopitz, Wärmeübertragung: Grundlagen, Analytische und Numerische Methoden, 1. ed., Pearson Studium, München, 2005.

78. L. de Quay, Validated Prediction of Pressurant Gas Requirements in Cryogenic Run Tanks at Subcritical and Supercritical Pressures: Dissertation, Mississippi State University, Mississippi, ProQuest LLC, Ann Arbor, Michigan, 2009.

79. L. de Quay and B.K. Hodge, A History of Collapse Factor Modeling and Empirical Data for Cryogenic Propellant Tanks, AIAA 2010-6559, in 46th AIAA/ASME/SAE/ASEE Joint Propulsion Conference & Exhibit, Nashville, Tennessee, USA, 2010.

80. E. Ring, Rocket Propellant and Pressurization Systems, Prentice-Hall, Englewood Cliffs and N. J., 1964.

81. W.H. Roudebush, An Analysis of the Problem of Tank Pressurization during Outflow, NASA TN D-2585, 1965.

82. W.H. Roudebush and D.A Mandell, Analytical Investigation of Some Important Parameters in the Pressurized Liquid Hydrogen Tank Outflow Problem, NASA TM X-52074, 1965.

83. A. Royon-Lebeaud, E.J. Hopfinger and A. Cartellier, Liquid Sloshing and Wave Breaking in Circular and Square-Base Cylindrical Containers, Journal of Fluid Mechanics 577, 2007, 467–494.

84. G.P. Sasmal, J.I. Hochstein and T.L. Hardy, Influence of Heat Transfer Rates on Pressurization of Liquid/Slush Hydrogen Propellant Tanks, in 31st Aerospace Sciences Meeting & Exhibit, Reno, Nevada, 1993.

85. G.P. Sasmal, J.I. Hochstein, M.C. Wendl and T.L. Hardy, Computational Modeling of the Pressurization Process in a NASP Vehicle Propellant Tank Experimental Simulation, AIAA 91-2407, AIAA Joint Propulsion Conference and Exhibit, 1991.

86. M.P. Segel, Experimental Study of the Phenomena of Stratification and Pressurization of Liquid Hydrogen, in International Advances in Cryogenic Engineering, Plenum Press, New York (1965), 308–313.

87. P.B. de Selding, European Space Agency Ministerial Backs Ariane 5 ME, Punts on Ariane 6, in Space News, 26.11.2012, vol. 23, iss. 46, 1-6.

88. M. Seo and S. Jeong, Analysis of Self-Pressurization Phenomenon of Cryogenic Fluid Storage Tank with Thermal Diffusion Model, Cryogenics, (50), 2010, 549–555.

89. R. Sionkiewicz and M. Dreyer, Experimental Analysis of Heat Transfer in Cryogenic Liquids, Technical Report for MT Aerospace, 16.03.2010.

90. R.J. Stochl, J.E. Maloy, P.A. Masters and R.L. DeWitt, Gaseous-Helium Requirements for the Discharge of Liquid Hydrogen from a 1.52-Meter- (5-ft-) Diameter Spherical Tank, NASA TN D-5621, 1970.

91. R.J. Stochl, J.E. Maloy, P.A. Masters and R.L. DeWitt, Gaseous-Helium Requirements for the Discharge of Liquid Hydrogen from a 3.96-Meter- (13-ft-) Diameter Spherical Tank, NASA TN D-7019, 1970.

92. R.J. Stochl, P.A. Masters, R.L. DeWitt and J.E. Maloy, Gaseous-Hydrogen Requirements for the Discharge of Liquid Hydrogen from a 1.52-Meter- (5-ft-) Diameter Spherical Tank, NASA TN D-5336, 1969.

93. R.J. Stochl, P.A. Masters, R.L. DeWitt and J.E. Maloy, Gaseous-Hydrogen Requirements for the Discharge of Liquid Hydrogen from a 3.96-Meter- (13-ft-) Diameter Spherical Tank, NASA TN D-5387, 1969.

94. G.P. Sutton, History of Liquid Propellant Rocket Engines: Library of Flight Series, AIAA, Reston and Viginia, 2006.

95. G.P. Sutton and O. Biblarz, Rocket Propulsion Elements, 8. ed., Wiley, Hoboken and NJ, 2010.

96. J.F. Thompson and M.E. Nein, Prediction of Propellant Tank Pressurization Requirements by Dimensional Analysis, NASA TN D-3451, 1966.

97. G.J. van Wylen, S.K. Fenster, H. JR. Merte and W.A. Warren, Pressurized Discharge of Liquid Nitrogen from an Uninsulated Tank, in Advances in Cryogenic Engineering, vol. 4, Plenum Press, New York (1960), 395–402.

98. M. Wade, Encyclopedia Astronautica: LOX/LH2, 18.01.2013.

99. L. Wang, Y. Li, Z. Zhao and Z. Liu, Transient Thermal and Pressurization Performance of LO2 Tank during Helium Pressurization combined with Outside Aerodynamic Heating, International Journal of Heat and Mass Transfer 62, 2013, 263–271.

100. L. Wang, Y. Li, Z. Zhao and J. Zheng, Numerical Investigation of Pressurization Performance in Cryogenic Tank of New-Style Launch Vehicle, Asia-Pacific Journal of Chemical Engineering, doi: 10.1002/apj.1746.

101. L. Wang, Y. Li, C. Li and Z. Zhao, CFD Investigation of Thermal and Pressurization Performance in LH2 Tank during Discharge, Cryogenics (57), 2013, 63–73.

102. F.H. Wenz, Die legendäre EUROPA Trägerrakete: Geschichte und Technik der in Deutschland gebauten 3. Stufe: 1961-1973, 1. ed., Stedinger, Lemwerder, 2004.

103. F.M. White, Fluid Mechanics, 4. ed., McGraw-Hill Science/Engineering/Math, 1998.

In the presented study, results are included which are generated in the context of the supervision of the following student work:

In der vorliegenden Arbeit sind Ergebnisse enthalten, die im Rahmen der Betreuung folgender studentischen Arbeiten entstanden sind:

- P. Friese, Charakterisierung und Thermalanalyse eines Teststandes zur aktiven Bedrückung von Tanks für kryogene Flüssigkeiten, Bachelorarbeit, Universität Bremen.

Appendix

Table A.1 Positions of the temperature sensors inside the tank. The error is $\pm\ 0.5{\cdot}10^{-3}$ m.

sensor	T1	T2	T3	T4	T5	T6	T7	T8
r [m]	0.0980	0.0980	0.0980	0.0980	0.0980	0.0980	0.0980	0.0980
z [m]	0.330	0.430	0.440	0.450	0.460	0.510	0.610	0.110
z/R	2.230	2.905	2.973	3.041	3.108	3.446	4.122	0.743

sensor	T9	T10	T11	T12	T13	T14	T15	T16
r [m]	-0.1480	-0.1480	-0.1480	-0.1480	-0.0980	0.0063	0.0980	0.0980
z [m]	0.450	0.460	0.625	0.630	0.650	0.644	0.435	0.445
z/R	3.041	3.108	4.223	4.257	4.392	4.351	2.939	3.007

Table A.2 Overview of the measurement equipment for the experiments with the corresponding typical measurement errors.

name	manufacturer	type	measurement range	serial number	$\pm$ typ. error
T1 - T16	LakeShore	DT-670 Band-B	1.4 - 500 K	-	0.1 K
T17	Jumo	Class B	73 - 273 K	-	2.1 K
			273 - 373 K	-	0.3 K
p_1	Sensor-technics	CTE9010ANO	$0 - 10^3$ kPa	2404	7.4 kPa
p_2	Honeywell	TJE	$0 - 5 \cdot 10^2$ kPa	1256143	3.2 kPa
p_3	Oerlikon Leybold Vacuum	Penningvac PTR90	$5 \cdot 10^{-10}$ kPa - 10^2 kPa	230071	$175{\cdot}10^{-9}$ kPa
m1	Bronkhorst	F-201-AV-70K-ABD-99V	$0 - 512.8{\cdot}10^{-6}$ kg/s	M9208488A	$13.9{\cdot}10^{-6}$ kg/s
m2	Bronkhorst	F-201C-FB-22V	$0 - 216{\cdot}10^{-6}$ kg/s	960945BA	$4.5{\cdot}10^{-6}$ kg/s

Table A.3 Fluid properties nitrogen, hydrogen, oxygen and helium at 101.3 kPa.

	T_{sat}	ρ	μ 10^{-6}	c_p	λ	Δh_v 10^5
	[K]	[kg/m^3]	[Pa s]	[J/(kg K)]	[W/(m K)]	[J/kg]
LN2	77.35	806.09	160.67	2041.5	0.146	1.99
LH2	20.37	70.85	13.32	9771.7	0.104	4.45
LOX	90.16	1141.20	195.83	1699	0.152	2.13
GN2	77.35	4.61	5.44	1124	0.008	
GH2	20.37	1.33	1.08	12036	0.017	
GOX	90.16	4.47	7.01	970	0.008	
GHe	20.37	2.39	3.62	5248.6	0.027	
	77.35	0.63	8.33	5195.7	0.062	
	90.16	0.54	9.16	5194.8	0.069	

Table A.4 GHe pressurized experiments: GN2 masses and partial pressures at pressurization end (index "f") and relaxation end (index "T").

Exp.	$m_{GN2,f}$ [kg]	$m_{GN2,T}$ [kg]	$p_{GN2,f}$ [kPa]	$p_{GHe,f}$ [kPa]	$p_{GN2,T}$ [kPa]	$p_{GHe,T}$ [kPa]
He200c	0.0423	0.0343	144.47	56.73	112.65	54.49
He400c	0.0437	0.0343	160.95	239.50	117.45	222.94
He200h	0.0390	0.0336	138.90	61.20	110.54	56.50
He400h	0.0528	0.0403	190.10	210.95	140.11	203.31

Table A.5 GHe pressurized experiments: Mass fractions Ψ and molar fractions χ at pressurization end (index "f") and relaxation end (index "T").

Exp.	$\Psi_{GN2,f}$ [ ]	$\Psi_{GHe,f}$ [ ]	$\Psi_{GN2,T}$ [ ]	$\Psi_{GHe,T}$ [ ]	$\chi_{GN2,f}$ [ ]	$\chi_{GHe,f}$ [ ]	$\chi_{GN2,T}$ [ ]	$\chi_{GHe,T}$ [ ]
He200c	0.95	0.05	0.94	0.06	0.72	0.28	0.33	0.67
He400c	0.82	0.18	0.79	0.21	0.40	0.60	0.35	0.65
He200h	0.94	0.06	0.93	0.07	0.69	0.31	0.66	0.34
He400h	0.86	0.14	0.83	0.17	0.47	0.53	0.41	0.59

Table A.6 Pressure plots are smoothened with a local regression using weighted linear least squares and a second order degree polynomial model. In the table, the coefficients of the polynom $p(x) = p_0 + p_1 x + p_2 x^2$ are summarized together with the span (specified as percentage of the total number of data points in the data set, e.g. span $= 0.1$ uses 10 % of the data points). Smoothing is performed after nondimensionalization. The data of N300a_short is used for Figures 6.1, 6.8 and 6.26 and N300a_long is used for Figures 6.24 and A.3.

Exp.	span	p_0	p_1	p_2
N300r	0.002	1.8648	1534.5	$-1.0163 \cdot 10^6$
N200c	0.002	1.5745	413.5969	$-1.7509 \cdot 10^5$
N300c	0.002	2.1308	746.8393	$-2.69315 \cdot 10^5$
N400c	0.002	3.026	251.6831	$-4.6041 \cdot 10^4$
N300a	0.002	1.5839	66807.0	$-1.5126 \cdot 10^7$
N300h_short	0.01	1.6595	2917.9	$-2.0666 \cdot 10^6$
N300h_long	0.01	1.6595	0.0575	$-8.0173 \cdot 10^4$
N300aH	0.01	2.2659	-0.014	$1.8411 \cdot 10^{-4}$
He200c	0.002	1.2754	2880.9	$-4.0764 \cdot 10^6$
He400c	0.002	2.6216	2494.1	$-1.6285 \cdot 10^6$

Table A.7 N300aH experiment: Theoretical and nondimensional temperature data for the stratification analysis. T_{sat} at $z = 0.455$ m is calculated with the NIST database [62].

z	z/R	T at $t_{p,0}$	$t_{p,f}$	$t_{p,T}$	T^* at $t_{p,0}$	$t_{p,f}$	$t_{p,T}$
[m]	[ ]	[K]	[K]	[K]	[ ]	[ ]	[ ]
0.650	4.392	281.52		281.65	0.743		0.744
0.644	4.351	275.25		281.22	0.721		0.742
0.610	4.122	239.69		246.03	0.591		0.614
0.510	3.446	135.11		142.51	0.210		0.237
0.460	3.108	82.82		90.76	0.020		0.049
0.455	3.074	77.60	87.94	85.58	0.009	0.039	0.030
0.455	3.074	77.60	87.94	85.58	0.024	1.000	0.780
0.45375	3.066		84.39			0.667	
0.4525	3.0574		81.48			0.391	
0.4524	3.0568		82.90			0.526	
0.45125	3.049	79.50			0.204		
0.450	3.041		78.39	80.69		0.099	0.317
0.4475	3.024	77.68	79.16		0.0313	0.172	
0.445	3.007		77.61	78.97		0.025	0.154
0.4425	2.990		77.88			0.050	
0.440	2.973		77.60	77.73		0.024	0.036
0.435	2.939		77.60	77.66		0.024	0.029
0.430	2.905		77.60	77.66		0.024	0.029
0.330	2.230		77.60	77.66		0.024	0.029
0.110	0.743		77.60	77.66		0.024	0.029

Table A.8 Experimental and nondimensional temperature and pressure data and relevant times for all performed experiments. T_{pg} is measured at temperature sensor T17. There is no data for the relaxation phase of the N400r experiment. The maximal error for T_{pg} is $\pm$ 0.3 K, except for $T_{pg} = 144$ K with an error of $\pm$ 1.5 K.

Exp.	T_{pg}	T_{pg}^*	p_0	p_f	p_f^*	p_m	p_T	p2	$t_{p,0}$ 10^{-4}	$t_{p,0}^*$	$t_{p,f}$ 10^{-4}	$t_{p,f}^*$	$t_{p,T}$ 10^{-4}	$t_{p,T}^*$	t_{press}	t_{relax}	Δp	Δp^* Flow $-$ 3D	t_{press}
	[K]	[ ]	[kPa]	[kPa]	[ ]	[kPa]	[kPa]	[kPa]	[s]	[ ]	[s]	[ ]	[s]	[ ]	[s]	[s]	[kPa]	[ ]	[s]
N200r	144	0.24	107.3	200.4	1.89	200.7	169.9	182	5.0	0.006	32.0	0.41	91.9	1.18	27.0	59.9	30.76	0.29	40.0
N300r	144	0.24	108.3	300.2	2.84	300.2	241.4	232	5.0	0.006	84.9	1.09	416.2	5.32	79.9	331.3	58.83	0.56	85.0
N400r	144	0.24	107.0	400.1	3.78	400.1	-	282	5.0	-	160.1	-	-	-	155.2	-	-	-	130.0
N200c	263	0.68	106.5	200.2	1.89	203.6	166.8	176	0.0	0.00	24.1	0.31	79.6	1.02	24.1	55.5	36.76	0.35	26.0
N300c	263	0.68	105.0	302.1	2.85	308.9	252.9	230	5.0	0.006	65.9	0.84	244.8	3.13	60.9	178.9	55.94	0.53	62.0
N400c	263	0.68	107.9	402.5	3.80	404.0	331.6	296	5.0	0.006	101.8	1.30	342.8	4.38	96.8	241.0	72.31	0.68	97.0
N200a	294	0.79	103.2	200.6	1.90	201.0	165.8	182	5.0	0.006	30.5	0.39	85.7	1.10	25.5	55.2	35.16	0.33	26.0
N300a	294	0.79	103.3	300.4	2.84	303.1	248.9	233	5.0	0.006	69.3	0.89	217.6	2.78	64.3	148.3	54.13	0.51	62.0
N400a	294	0.79	106.4	399.9	3.78	399.9	329.6	282	5.0	0.006	105.1	1.34	357.0	4.65	100.1	251.9	70.21	0.66	102.0
N200h	352	1.00	104.5	198.0	1.87	204.2	165.8	186	5.0	0.006	28.6	0.37	92.7	1.19	23.6	64.1	38.35	0.36	25.0
N300h	352	1.00	104.2	300.1	2.84	301.7	245.4	235	5.0	0.006	65.7	0.84	218.0	2.79	60.7	152.3	56.24	0.53	60.0
N400h	352	1.00	107.2	402.3	3.80	406.2	333.5	285	5.0	0.006	101.8	1.30	355.9	4.55	96.8	254.1	72.61	0.69	96.0
N300aH	294	0.79	101.0	300.9	2.84	300.9	242.5	235	5.0	0.006	57.7	0.74	188.8	2.41	52.7	131.1	58.43	0.55	50.0
He200c	263	0.68	105.3	201.2	1.90	201.7	167.1	168	5.0	0.006	19.6	0.25	110.5	1.41	14.6	90.9	34.56	0.33	5.0
He400c	263	0.68	107.6	400.5	3.78	407.0	340.4	272	5.0	0.006	62.2	0.80	250.7	3.21	57.2	188.5	66.56	0.63	29.0
He200h	352	1.00	105.8	200.1	1.89	203.0	167.0	167	5.0	0.006	20.1	0.26	110.7	1.42	15.1	90.6	35.96	0.34	3.0
He400h	352	1.00	105.3	401.1	3.79	405.5	343.4	290	5.0	0.006	56.5	0.72	228.2	2.92	51.5	171.7	62.03	0.59	18.0

Table A.9 Dimensional and nondimensional masses for vapor and liquid phase, pressurant gas as well as condensed GN2. Specific enthalpies of pressurant gas and condensed GN2 from the NIST database with default reference states [62]. Specific enthalpy of pressurant gas determined with T_{pg} and p$_2$ from Table A.8.

Exp.	$m_{v,0}$	$m_{v,f}$	$m_{v,T}$	$m_{l,0}$	m_{pg}	h_{pg}	$m_{cond,0,f}$	$m_{cond,f,T}$	$h_{v,cond,0,f}$	$h_{v,cond,f,T}$	$m \cdot h$ pg	$m \cdot h$ cond,0,f	$m \cdot h$ cond,f,T	m_{pg}^*	$m_{v,0}^*$	$m_{cond,0,f}^*$	$m_{cond,f,T}^*$
						10^5			10^3	10^3	10^3	10^3	10^3				
	[kg]	[kg]	[kg]	[kg]	[kg]	[J/kg]	[kg]	[kg]	[J/kg]	[J/kg]	[J]	[J]	[J]	[ ]	[ ]	[ ]	[ ]
N200r	0.0355	0.0566	0.0522	23.4075	0.0225	1.50	0.0014	0.0044	79.85	81.03	3.38	0.11	0.36	0.6420	1.0131	0.0393	0.1255
N300r	0.0359	0.0831	0.0720	23.4172	0.0665	1.49	0.0194	0.0110	81.65	83.36	9.91	1.58	0.92	1.8998	1.0269	0.5534	0.3156
N400r	0.0354	0.1112	-	23.4156	0.1292	1.44	0.0534	-	82.97	-	18.60	4.43	-	3.6878	1.0119	1.5257	-
N200c	0.0351	0.0543	0.0511	23.4158	0.0201	2.73	0.0009	0.0032	79.83	80.98	5.49	0.07	0.26	0.5730	1.0033	0.0259	0.0914
N300c	0.0346	0.0794	0.0738	23.4243	0.0507	2.72	0.0059	0.0056	81.63	83.51	13.79	0.48	0.47	1.4480	0.9883	0.1692	0.1600
N400c	0.0355	0.1037	0.0942	23.4187	0.0806	2.72	0.0123	0.0095	83.01	85.07	21.92	1.02	0.81	2.3017	1.0145	0.3525	0.2716
N200a	0.0340	0.0553	0.0505	23.4377	0.0212	3.05	0.0000	0.0047	79.77	80.97	4.47	0.00	0.38	0.6063	0.9727	0.0006	0.1350
N300a	0.0342	0.0786	0.0725	23.4206	0.0535	3.05	0.0091	0.0061	81.57	83.45	16.32	0.74	0.51	1.5288	0.9772	0.2605	0.1739
N400a	0.0355	0.1038	0.0942	23.4135	0.0833	3.05	0.0150	0.0096	82.96	85.03	25.41	1.24	0.82	2.3801	1.0152	0.4293	0.2742
N200h	0.0344	0.0542	0.0505	23.4217	0.0196	3.65	-0.0001	0.0037	79.74	80.96	7.15	-0.01	0.30	0.5611	0.9837	-0.0035	0.1060
N300h	0.0345	0.0787	0.0720	23.4199	0.0505	3.65	0.0064	0.0066	81.58	83.41	18.43	0.52	0.55	1.4433	0.9859	0.1815	0.1896
N400h	0.0355	0.1039	0.0952	23.3950	0.0806	3.65	0.0122	0.0087	83.00	85.08	29.42	1.01	0.74	2.3016	1.0134	0.3480	0.2486
N300aH	0.0319	0.0754	0.0683	23.9873	0.0439	3.05	0.0004	0.0071	81.55	80.98	16.02	0.03	0.57	1.3748	1.0000	0.0124	0.2015
He200c	0.0347	0.0446	0.0367	23.4067	0.0024	13.73	-0.0076	0.0080	78.51	78.70	3.30	-0.60	0.63	0.0677	0.9900	-0.2174	0.2274
He400c	0.0355	0.0530	0.0435	23.4141	0.0093	13.70	-0.0082	0.0094	78.98	79.21	12.74	-0.65	0.74	0.2654	1.0138	-0.2346	0.2697
He200h	0.0348	0.0414	0.0360	23.4144	0.0025	18.28	-0.0042	0.0054	78.38	78.50	4.57	-0.33	0.42	0.0701	0.9934	-0.1195	0.1536
He400h	0.0345	0.0611	0.0487	23.4107	0.0084	18.35	-0.0183	0.0124	79.59	80.30	15.41	-1.46	1.00	0.2390	0.9843	-0.5228	0.3546

Table A.10 Average temperatures and specific internal energies (determined from the NIST database [62]) at pressurization start (index "0"), end (index "f") and relaxation end (index "T"). Change in heat of the liquid and vapor phase as well as the total tank for the pressurization phase (index "$0, f$") and relaxation phase (index "f, T").

Exp.	$\bar{T}_{v,0}$	$\bar{T}_{v,f}$	$\bar{T}_{v,T}$	$\bar{T}_{l,0}$	$\bar{T}_{l,f}$	$\bar{T}_{l,T}$	$u_{v,0}$ 10^3	$u_{v,f}$ 10^3	$u_{v,T}$ 10^3	$u_{l,0}$ 10^3	$u_{l,f}$ 10^3	$u_{l,T}$ 10^3	$dQ_{v,0,f}$ 10^3	$dQ_{l,0,f}$ 10^3	$dQ_{v,f,T}$ 10^3	$dQ_{l,f,T}$ 10^3	$dQ_{0,f}$ 10^3	$dQ_{f,T}$ 10^3
	[K]	[K]	[K]	[K]	[K]	[K]	[J/kg]	[J/kg]	[J/kg]	[J/kg]	[J/kg]	[J/kg]	[J]	[J]	[J]	[J]	[J]	[J]
N200r	143.84	168.41	154.83	77.78	77.82	77.86	105.81	123.75	113.71	-121.28	-121.25	-121.15	-0.017	0.426	-0.711	1.453	0.409	0.742
N300r	143.24	171.73	159.25	77.70	77.88	78.14	105.36	125.79	116.66	-121.44	-121.18	-120.62	-0.166	2.160	-1.126	10.871	0.500	9.745
N400r	143.61	170.98	-	77.72	78.02	-	105.64	124.76	-	-121.40	-121.95	-	-3.983	-0.352	-	-	-4.335	-
N200c	144.16	171.08	156.08	77.72	77.80	77.81	106.06	125.76	114.66	-121.40	-121.29	-121.25	-2.294	2.393	-0.710	0.290	0.099	-0.420
N300c	144.30	180.91	162.98	77.65	77.80	77.96	106.17	132.70	119.41	-121.54	-121.35	-120.99	-6.448	3.249	-1.256	7.289	-3.200	6.034
N400c	144.45	184.37	167.26	77.69	77.93	78.22	106.27	134.89	122.26	-121.46	-121.14	-120.51	-10.680	4.976	-1.664	12.807	-5.704	11.443
N200a	144.10	172.57	156.01	77.55	77.63	77.65	106.03	126.87	114.61	-121.75	-121.64	-121.58	-3.079	2.574	-0.836	0.449	-0.505	-0.387
N300a	143.59	181.63	163.16	77.68	77.90	78.05	105.65	133.24	119.57	-121.48	-121.40	-120.81	-8.715	6.115	-1.294	6.488	-2.600	5.194
N400a	142.34	183.04	166.27	77.73	77.99	78.30	104.70	133.89	121.53	-121.38	-121.01	-120.34	-13.977	5.598	-1.634	13.726	-8.379	12.093
N200h	144.29	173.55	156.12	77.67	77.73	77.76	106.17	127.61	114.70	-121.50	-121.43	-121.35	-3.915	1.664	-0.829	1.123	-2.251	0.294
N300h	143.56	181.28	161.92	77.69	77.86	77.98	105.62	132.98	118.65	-121.46	-121.22	-120.95	-11.116	4.333	-1.361	4.969	-6.784	3.608
N400h	143.66	184.07	166.59	77.87	78.10	78.37	105.68	134.66	121.75	-121.10	-120.79	-120.20	-18.164	4.771	-1.660	12.024	-13.393	10.364
N300aH	143.17	180.48	160.44	77.58	77.69	77.67	105.35	132.38	117.55	-121.69	-121.57	-121.39	-6.714	2.798	-1.359	2.873	-3.916	1.515
He200c	144.47	162.48	156.07	77.78	77.82	77.88	106.30	140.39	139.29	-121.28	-121.22	-121.08	-1.271	2.924	-0.531	1.686	1.653	1.155
He400c	144.15	175.08	162.98	77.73	77.92	78.08	106.04	202.90	203.93	-121.38	-121.02	-120.67	-6.392	10.071	-1.223	6.305	3.680	5.182
He200h	144.65	169.48	156.47	77.73	77.82	77.85	106.43	149.04	140.95	-121.38	-121.22	-121.14	-2.342	4.581	-0.671	0.799	2.240	0.129
He400h	145.30	171.28	165.08	77.75	77.95	78.10	106.92	182.47	189.90	-121.34	-120.98	-120.64	-9.336	12.098	-0.906	5.460	2.762	4.553

Table A.11 N300aH experiment: Experimental and nondimensional temperature data for the stratification analysis. T_{sat} at $z = 0.455$ m is calculated with the NIST database [62].

z [m]	z/R [ ]	T at $t_{p,0}$ [K]	$t_{p,f}$ [K]	$t_{p,T}$ [K]	T^* at $t_{p,0}$ [ ]	$t_{p,f}$ [ ]	$t_{p,T}$ [ ]
0.650	4.392	281.52	281.70	281.63	0.742	0.742	0.743
0.644	4.351	273.82	292.41	281.75	0.715	0.783	0.744
0.610	4.122	229.96	266.59	247.36	0.556	0.689	0.619
0.510	3.446	127.53	177.00	149.83	0.183	0.363	0.264
0.460	3.108	79.81	101.73	91.49	0.009	0.089	0.051
0.455	3.074	77.60	87.94	85.60	0.001	0.039	0.030
0.455	3.074	77.60	87.94	85.60	0.024	1.000	0.782
0.450	3.041	77.56	78.31	80.67	0.020	0.091	0.323
0.445	3.007	77.67	77.82	78.79	0.030	0.045	0.136
0.440	2.973	77.55	77.64	77.91	0.019	0.028	0.053
0.435	2.939	77.64	77.72	77.79	0.027	0.035	0.042
0.430	2.905	77.58	77.65	77.68	0.022	0.028	0.031
0.330	2.230	77.62	77.64	77.69	0.026	0.028	0.032
0.110	0.743	77.56	77.60	77.64	0.020	0.024	0.027

Table A.12 N400h and He400h experiments: Experimental and nondimensional temperature data for the stratification analysis. T_{sat} at $z = 0.445$ m is calculated with the NIST database [62].

z [m]	z/R [ ]	T at N400h $t_{p,0}$ [K]	$t_{p,f}$ [K]	$t_{p,T}$ [K]	He400h $t_{p,0}$ [K]	$t_{p,f}$ [K]	$t_{p,T}$ [K]	T^* at N400h $t_{p,0}$ [ ]	$t_{p,f}$ [ ]	$t_{p,T}$ [ ]	He400h $t_{p,0}$ [ ]	$t_{p,f}$ [ ]	$t_{p,T}$ [ ]
0.650	4.392	276.23	277.62	277.16	280.42	281.39	280.96	0.724	0.729	0.728	0.739	0.743	0.741
0.644	4.351	270.56	323.52	279.09	272.81	325.89	284.76	0.703	0.896	0.735	0.712	0.905	0.755
0.610	4.122	230.32	274.73	250.27	233.27	269.97	248.28	0.557	0.719	0.630	0.568	0.701	0.622
0.510	3.446	134.71	183.76	163.03	136.20	192.73	175.32	0.209	0.387	0.312	0.214	0.420	0.357
0.460	3.108	87.02	119.95	108.06	87.98	135.79	110.54	0.035	0.155	0.112	0.039	0.213	0.121
0.450	3.041	79.24	101.91	94.53	79.39	109.78	94.54	0.007	0.089	0.063	0.007	0.118	0.063
0.445	3.007	77.83	91.30	89.10	77.69	91.26	89.44	0.002	0.051	0.043	0.001	0.051	0.044
0.445	3.007	77.83	91.30	89.10	77.69	91.26	89.44	0.035	1.000	0.848	0.025	1.000	0.873
0.440	2.973	77.84	80.48	84.40	77.70	80.20	82.45	0.035	0.226	0.509	0.025	0.206	0.368
0.430	2.905	77.92	78.09	79.59	77.72	77.90	78.79	0.041	0.053	0.162	0.027	0.040	0.104
0.330	2.230	77.88	77.96	78.00	77.76	77.87	77.91	0.038	0.044	0.047	0.030	0.038	0.040
0.110	0.743	77.86	77.95	77.97	77.76	77.87	77.87	0.037	0.043	0.045	0.030	0.038	0.038

Table A.13 Relevant data and results of Equation 6.53 for the nondimensional required pressurant gas mass of all performed experiments. The initial saturation temperature $T_{sat,p,0} = 77.35$ K, the constant K of Equation 6.53 is assumed as 1 and $m_{ref} = 1$ kg. The pressurant gas inlet velocity v_{pg} is calculated with pressurant gas inlet temperature at the final tank pressure p_f and errors are given of $\tilde{m}_{pg,cn}$ relative to $\tilde{m}_{pg,exp}$. All calculated data are from the NIST database [62].

Exp.	T_{pg}	p_f	V_u	$\tilde{m}_{pg,exp}$	$\frac{V_u}{A_\Gamma}$	$T_{sat,p,f}$	$c_{p,l}$ 10^3	$\Delta h_{v,p,f}$ 10^5	v_{pg}	$\beta_{T,pg}$	Ja	$\mathbf{Fr}_{\beta_T\mathbf{pg}}$	$\tilde{m}_{pg,cn}$	error	abs. error
	[K]	[kPa]	[%]	[ ]	[m]	[K]	[J/(kg K)]	[J/kg]	[m/s]	[1/K]	[ ]	[ ]	[ ]	[%]	[ ]
N200c	263	200	32.8	0.0201	0.205	83.63	2.08	1.91	1.00	0.0038	0.0685	0.7058	0.0201	0.00	0.0000
N300c	263	300	32.8	0.0507	0.205	87.91	2.12	1.84	0.67	0.0039	0.1215	0.3121	0.0624	23.15	0.0117
N400c	263	400	32.8	0.0806	0.205	91.23	2.16	1.78	0.50	0.0039	0.1678	0.1747	0.1230	52.55	0.0424
N200a	294	200	32.8	0.0212	0.205	83.63	2.08	1.91	1.12	0.0034	0.0685	0.8479	0.0189	-10.65	-0.0023
N300a	294	300	32.8	0.0535	0.205	87.91	2.12	1.84	0.75	0.0034	0.1215	0.3757	0.0587	9.71	0.0052
N400a	294	400	32.8	0.0833	0.205	91.23	2.16	1.78	0.56	0.0034	0.1678	0.2107	0.1155	38.66	0.0322
N200h	352	200	32.8	0.0196	0.205	83.63	2.08	1.91	1.34	0.0028	0.0685	1.1523	0.0171	-12.75	-0.0025
N300h	352	300	32.8	0.0505	0.205	87.91	2.12	1.84	0.89	0.0029	0.1215	0.5115	0.0530	4.86	0.0025
N400h	352	400	32.8	0.0806	0.205	91.23	2.16	1.78	0.67	0.0029	0.1678	0.2874	0.1042	29.23	0.0236
N200r	144	200	32.8	0.0255	0.205	83.63	2.08	1.91	0.54	0.0072	0.0685	0.3046	0.0267	4.49	0.0011
N300r	144	300	32.8	0.0655	0.205	87.91	2.12	1.84	0.36	0.0074	0.1215	0.1308	0.0834	27.38	0.0179
N400r	144	400	32.8	0.1292	0.205	91.23	2.16	1.78	0.27	0.0075	0.1678	0.0710	0.1660	28.48	0.0368
N300aH	294	300	31.2	0.0439	0.195	87.91	2.12	1.84	0.75	0.0034	0.1215	0.4076	0.0571	30.10	0.0132
He200c	263	200	32.8	0.0024	0.205	83.63	2.08	1.91	1.37	0.0038	0.0685	1.3330	0.0163	578.76	0.0139
He400c	263	400	32.8	0.0093	0.205	91.23	2.16	1.78	0.69	0.0038	0.1678	0.3344	0.0990	964.84	0.0897
He200h	352	200	32.8	0.0025	0.205	83.63	2.08	1.91	1.83	0.0028	0.0685	2.1585	0.01387	454.90	0.0114
He400h	352	200	32.8	0.0084	0.205	91.23	2.16	1.78	0.92	0.0028	0.1678	0.5409	0.0844	904.31	0.0760

Table A.14 Relevant data of experiments by van Dresar and Stochl [34] and results of Equation 6.53 for the nondimensional required pressurant gas mass of experiments #487 to #494 and for experiments #497 to #505 with Equation 6.54. The injector area is $A_{dif} = 0.057$ m^2. The average initial saturation temperature $T_{sat,p,0} = 20.27$ K, thermal expansion coefficient $\beta_{T,pg} = 0.0034$ 1/K, the constant K of Equations 6.53 and 6.54 is assumed as 1, $m_{ref} = 1$ kg and errors are given of $\tilde{m}_{pg,cn}$ relative to $\tilde{m}_{pg,exp}$. All calculated data are from the NIST database [62].

$Exp.$	T_{pg}	p_f	V_u	t_{press}	$\tilde{m}_{pg,exp}$	ρ_{pg}	$\dfrac{V_u}{A_\Gamma}$	$T_{sat,p,f}$	$c_{p,l}$ 10^4	$\Delta h_{v,p,f}$ 10^5	v_{pg}	**Ja**	**Fr**$_{\beta_T}$**pg**	$\tilde{m}_{pg,cn}$	error	abs. error
	[K]	[kPa]	[%]	[s]	[ ]	[kg/m^3]	[m]	[K]	[J/(kg K)]	[J/kg]	[m/s]	[ ]	[ ]	[ ]	[%]	[ ]
#488	271	207	13.0	14	0.0653	0.18496	0.6357	22.94	1.16	4.28	0.442	0.0735	0.0368	0.0600	-8.12	-0.0053
#487	268	203	13.0	24	0.0793	0.18342	0.6357	22.86	1.15	4.29	0.316	0.0708	0.0190	0.0706	-10.97	-0.0087
#489	272	205	13.0	57	0.1041	0.18250	0.6357	22.90	1.16	4.28	0.176	0.0722	0.0058	0.1081	3.84	0.0040
#490	273	204	14.0	108	0.1222	0.18095	0.2458	22.88	1.15	4.29	0.110	0.0715	0.0058	0.1063	-13.01	-0.0159
#494	284	278	14.0	24	0.1203	0.23694	0.2458	24.22	1.27	4.15	0.371	0.1225	0.0637	0.1073	-10.81	-0.0130
#493	284	278	13.0	39	0.1414	0.23694	0.6357	24.22	1.27	4.15	0.269	0.1225	0.0129	0.1829	29.35	0.0415
#492	282	277	13.0	92	0.1812	0.23776	0.6357	24.20	1.27	4.15	0.145	0.1218	0.0038	0.2722	50.22	0.0910
#491	280	275	13.0	180	0.2173	0.23773	0.6357	24.17	1.27	4.15	0.089	0.1204	0.0014	0.3699	70.23	0.1526
#498	290	203	50.0	61	0.2607	0.16951	0.6430	22.86	1.15	4.29	0.442	0.0708	0.0338	0.2190	-15.99	-0.0417
#497	289	203	50.0	116	0.3151	0.17010	0.6430	22.86	1.15	4.29	0.280	0.0708	0.0136	0.2388	-24.23	-0.0763
#505	291	203	51.0	294	0.4141	0.16893	0.6575	22.86	1.15	4.29	0.146	0.0708	0.0036	0.4618	11.52	0.0477
#499	292	272	50.0	98	0.4825	0.22548	0.6430	24.12	1.26	4.16	0.383	0.1153	0.0252	0.3935	-18.46	-0.0891
#500	292	276	50.0	188	0.5685	0.22879	0.6430	24.19	1.27	4.15	0.232	0.1211	0.0092	0.5775	1.58	0.0091
#501	291	276	51.0	493	0.6255	0.22958	0.6575	24.19	1.27	4.15	0.097	0.1211	0.0016	1.0390	66.11	0.4135

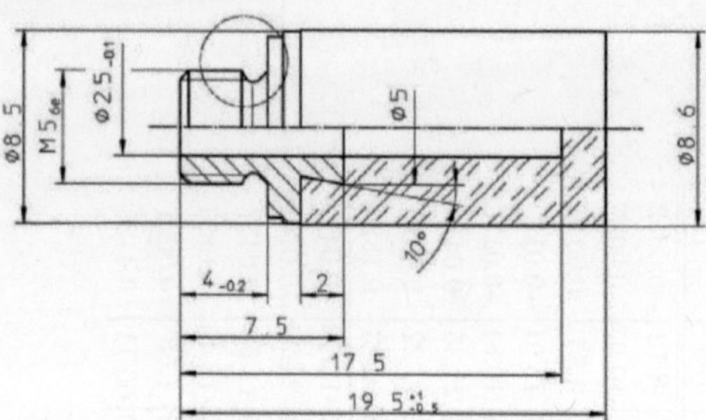

Fig. A.1 Design drawing of the Festo U-M5 sintered filter used as diffuser for the performed experiments [43] (dimensions in millimeter).

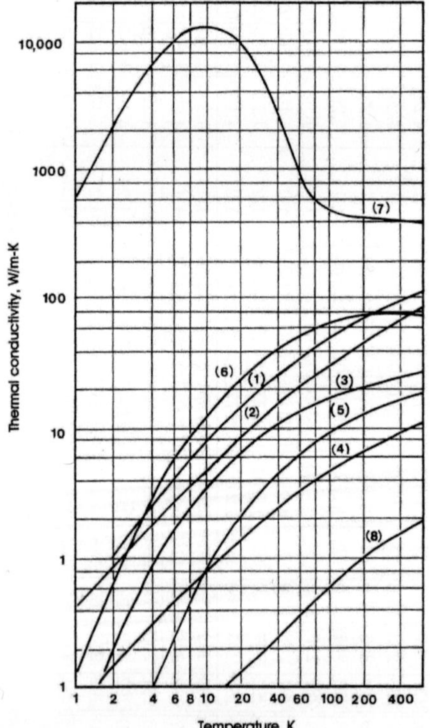

Fig. A.2 Thermal conductivity of solids at low temperatures: (1) 2024-T4 aluminum; (2) beryllium copper; (3) K Monel; (4) titanium; (5) 304 stainless steel; (6) C1020 carbon steel; (7) pure copper, single crystal; (8) Teflon (Stewart and Johnson 1961) from [11].

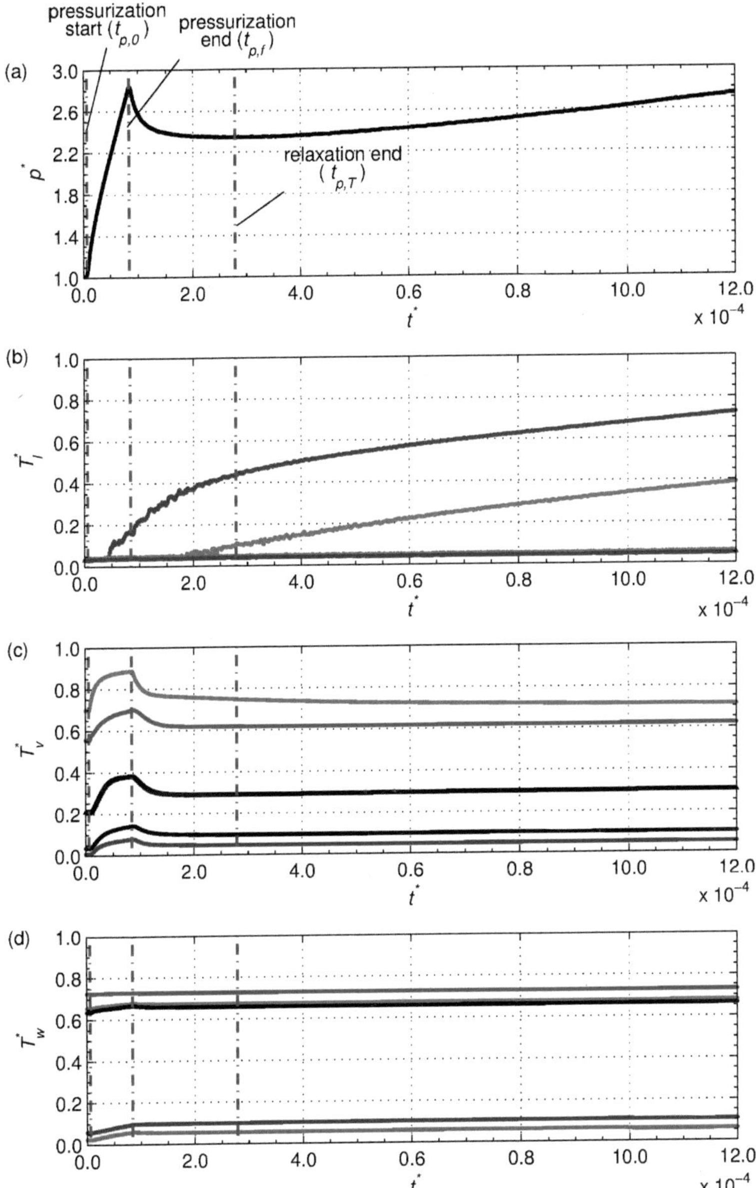

Fig. A.3 N300h experiment: nondimensional (a) tank pressure, (b) liquid temperatures, (c) vapor temperatures, (d) wall and lid temperatures until experiment end. The section from pressurization start until relaxation end is depicted more detailed in Figure 6.1.